Javier Nogueras-Iso
F. Javier Zarazaga-Soria
Pedro R. Muro-Medrano

Geographic Information Metadata for Spatial Data Infrastructures

Resources, Interoperability and Information Retrieval

T0142871

Javier Nogueras-Iso
F. Javier Zarazaga-Soria
Pedro R. Muro-Medrano

Geographic Information Metadata for Spatial Data Infrastructures

Resources, Interoperability and Information Retrieval

With 83 Figures

 Springer

Dr. Javier Nogueras-Iso
Dr. F. Javier Zarazaga-Soria
Dr. Pedro R. Muro-Medrano
University of Zaragoza
María de Luna, 1
50018 Zaragoza
Spain

E-mail: jnog@unizar.es
        javy@unizar.es
        prmuro@unizar.es

ISBN-13 978-3-642-06380-0   e-ISBN-13 978-3-540-27508-4

Springer is a part of Springer Science+Business Media
springeronline.com
© Springer-Verlag Berlin Heidelberg 2010
Printed in The Netherlands

Cover design: E. Kirchner, Heidelberg

Printed on acid-free paper 30/2132/AO 5 4 3 2 1 0

To the IAAA staff

To our families and friends

Preface

The Geographic Information (GI), also known as geo-spatial data, is the information that describes phenomena associated directly or indirectly with a location with respect to the Earth surface. Nowadays, there are available large amounts of geographic data that have been gathered (for decades) with different purposes by different institutions and companies. For instance, the geographic information is vital for decision-making and resource management in diverse areas (natural resources, facilities, cadastres, economy...), and at different levels (local, regional, national or even global) (Buehler and McKee, 1996). Furthermore, the volume of this information grows day by day thanks to important technology advances in high-resolution satellite remote sensors, Global Positioning Systems (GPS), databases and geo-processing software notwithstanding an increasing interest by individuals and institutions. Even more, it is possible to georeference complex collections of a broad range of resource types, including textual and graphic documents, digital geospatial map and imagery data, real-time acquired observations, legacy databases of tabular historical records, multimedia components such as audio and video, and scientific algorithms.

In recent years nations have made unprecedented investments in both information and the means to assemble, store, process, analyze, and disseminate it. Thousands of organizations and agencies (all levels of government, the private and non-profit sectors, and academia) throughout the world spend billions of euros each year producing and using geographic data (Somers, 1997; Groot and McLaughlin, 2000). This has been particularly enhanced by the rapid advancement in spatial data capture technologies, which has made the capture of digital spatial data a relatively quick and easy process. Additionally, it is also worthwhile mentioning the impact of the Internet in the distribution of geographic information resources. As well as other information resources, lots of geographic information resources are also available on the Internet. And in some cases it is even assumed that the own Internet is the storehouse of this information.

However, almost every new project or study implying the use of geographic information requires the creation of new geographic information resources from scratch. This apparent lack of reusable resources may be motivated by the following circumstances:

- Most organizations need more data than they can afford. It must be realized that the creation of geographic information requires in most cases important financial resources. For instance, the creation of topographic maps must include the financial support for aerial flights, topographers field sessions, apparatus and human resources for digitalization. Additionally, the high volumes of geographic information, e.g. raster data obtained by remote sensors, usually require high-density storage devices as well as well-organized backup and recovery policies. Last, geographic information is updated quite frequently. This originates problems of maintenance and control of the different version of this geographic information.
- Some organizations, despite being public institutions, are reticent to distribute high-quality information. Organizations often need data outside their jurisdictions or operational areas. However, the accessibility to the data is very limited because public administrations rarely have permission to facilitate the reuse of the data that was obtained for the particular need of these administrations. Besides, they usually lack the infrastructure to enable this reuse. Although they were willing to facilitate these data, there usually is neither political support nor the strategy to do it. Additionally, it may occur that the information needed to solve cross-jurisdictional problems (e.g., information needed for natural risk management systems in cross-border areas) does not exist.
- Data collected by different organizations are often incompatible. On one hand, sometimes geographic information producers do not take into account the multidisciplinary use of geographic information and create resources that are not general enough to be reused in other application domains. And on the other hand, the proliferation of exchange formats and their characterization also hinders the compatibility. During last decades almost each Geographic Information System (GIS) vendor has created its own specific formats to maximize the possibilities of its software. However, this implies interoperability problems when data are exchanged between two different geographic information system products. Geographic information system vendors have tried to overcome this problem by providing import/export utilities to enable compatibility. But this is not a seamless solution because these data conversions usually involve an information loss.
- In most cases, there is a lack of knowledge about what data is currently available. It is not unusual to find, that different divisions of the same company pay data suppliers for a product that had been already ordered by another division. This lack of synchronism leads into a consecutive recreation of data with similar characteristics.

- The poor quality and poor documentation of data that is available through the WWW. As it is mentioned in (U.S. National Research Council, 1999), the use of the WWW as a mechanism for storing and disseminating geoinformation has also introduced potential disadvantages. Little of the information now available via the WWW has been subjected to the mechanisms that ensure quality in traditional publication and library acquisition: peer review, editing, and proofreading. There are no WWW equivalents of the library's collection specialists who monitor library content. As the volume of information grows, issues of quality and reliability are becoming more complex. Problems of context, provenance and timeliness become much more complex with the added dimension of distribution. But it is easy to be misled into believing that quality control problems of the WWW are somehow different from conventional ones. Users of on-line digital geographic information will tend to trust data that come from reputable institutions, with documented assurances of quality, and to mistrust data of uncertain origins, just as they do today by acquiring them off-line. The problem is that, as mentioned before, many public administrations are still reticent to the distribution of geographic information resources.
- Another issue related with the use of Internet is the increasing complexity of discovery and information retrieval services. There is an increasing volume, diversity, decentralization and autonomy in the development, meaning and types of information. The number of protocols for accessing this information increases and the reasons for making it available are more complex than simply sharing useful data. At the same time, there is a massive growth in the number and diversity of users' sophistication and background, and expectations. There is also an increasing criticality of the search problem to people's personal and professional lives. Furthermore, not only human users are searching on the Web. At present, there are computing systems whose functionality is based on the discovered information, e.g. decision-support systems.

In conclusion, despite the potential uses of geographic information and the important investments in their creation, nowadays geographic information is not exploited enough. Several studies (Craglia et al., 1999; European Commission, 1998; Official Journal of the European Union, 2003) have remarked that although the value of geospatial data is recognized by both government and society, the effective use of geospatial data is inhibited by poor knowledge of the existence of data, poorly documented information about the data sets, and data inconsistencies. It is said that "information is power", but with increasing amounts of data being created and stored (but often not well organized) there is a real need to document the data for future use - to be as accessible as possible to as wide a "public" as possible. Data plus the context for its use (documentation) become information. Data without context are not as valuable as documented data. This necessity has an extremely importance in the case of geographic information. Once created, geospatial data can be used by multiple software systems for different purposes. Over thirty five years ago, humans landed on the Moon. Data from that era are still being used today, and it is reasonable to assume that today's geospatial data could still be used in the year 2020 and beyond to study climate change, ecosystems, and other natural processes.

As it can be deduced, there is a need for creating networked solutions to facilitate the discovery, evaluation and access to geographic data. A Spatial Data Infrastructure (SDI) is defined as the infrastructure that provides the framework for the optimization of the creation, maintenance and distribution of geographic information at different organization levels (e.g., regional, national, or global level) and involving both public and private institutions (Nebert, 2001). Governments start considering spatial data infrastructures as basic infrastructures for the development of a country. Indeed, they are becoming as relevant as classical infrastructures like utilities (water, electricity, gas), transport or telecommunication infrastructures. In this sense, it is worthwhile mentioning two high-level political decisions, among others, that have encouraged the development of spatial data infrastructures. On one hand, in April 1994 the U.S. president Bill Clinton signed an Executive Order (U.S. Federal Register, 1994) for the establishment of the "National Spatial Data Infrastructure" (NSDI), forcing the cooperation among federal and local agencies in collecting, spreading and using geographic information. And on the other hand, in November 2001 the European Commission launched INSPIRE (INfrastructure for SPatial InfoRmation in Europe), an initiative to create a European legislation to guide national and regional spatial data infrastructure development. This initiative, sponsored at the highest levels within the European Commission, will mandate how and when each member state should create its national spatial data infrastructure. The overall objective of this initiative is to enable the availability of a European spatial data infrastructure, which will consist in the cooperation of the different national and regional SDIs.

Because the concept of spatial data infrastructures comes from the geographic information domain, in many cases, they are being built over the concepts and experiences provided by the traditional geographic information systems with small references to other disciplines. Maybe the most relevant example can be found in the proposals made by the Open Geospatial Consortium Inc. (OGC)[1], which was created in 1994 under the name Open GIS Consortium as a member-driven, non-profit international trade association [2]. The

---

[1] http://www.opengeospatial.org/

[2] Up to April 2004 it integrates more than 250 members which include: leading companies in the GIS sector like ESRI (http://www.esri.com/), Intergraph (http://www.intergraph.com) or MapInfo (http://www.mapinfo.com); some of the main developers of hardware and software, e.g. Sun Microsystems (http://www.sun.com) or Oracle (http://www.oracle.com/); other relevant companies in different sectors like telecommunications or consultancy; governmental agencies involved in geoprocessing such as the U.S. National

vision of OGC is that of a world in which everyone benefits from geographic
information and services made available across any network, application, or
platform. Its mission is to promote the development and use of advanced open
systems standards and techniques in the area of geo-processing and related
information technologies delivering spatial interface specifications that are
openly available for global use. For that purpose, this consortium encourages
the creation of interoperability programs consisting of test-beds (collaborative
research and development efforts) and pilot projects (implementing specifications to serve real world applications) that accelerate the development and
testing of interfaces for plug-and-play software components enabling the geographic information interchange. Most of the specifications provided by OGC,
which have a geographic information systems slant, are being used in the development of spatial data infrastructures [3]. However, these approximations
could be improved by taken into account the experiences gain in other disciplines. In particular, Digital Libraries could contribute with a very important
background. They have a vast experience in technology for the distribution of
digital resources that could be used as the base for the development of spatial
data infrastructures own concepts, processes and methods.

One of the essential pieces in the development of spatial data infrastructures and digital libraries is the appropriate documentation of data and services. This documentation is called metadata and it is commonly defined as
"structured data about data" or "data which describes attributes of a resource" or, more simply, "information about data". Metadata offers description of the content, quality, condition, authorship and any other characteristics of the resource. It constitutes the mechanism to characterize data and
services in order to enable other users and applications to make use of such
data and services. Metadata records, each one describing a specific resource,
are usually published through catalog systems, sometimes also called directories or registries. Electronic catalogs do not differ very much from traditional
library catalogs (enumerating the resources of a library) except for the fact
that they offer a standardized interface of discovery services, which provide
users and applications with the possibility of finding the resources of their
interest. Thus, metadata and catalogs are the basic components that facilitate the accessibility and interoperability of the resources and services offered
by a spatial data infrastructure. Furthermore, the improvement in the use of

---

Aeronautics and Space Administration (NASA, http://www.nasa.gov/), the
U.S. Geological Survey (http://www.usgs.gov/), the European Union Satellite Centre (http://www.weusc.es/) or the Environment Department of the
Galicia Government(Xunta de Galicia - Consellería de Medio Ambiente,
http://www.xunta.es/conselle/cma/index.htm); and a large number of univer-
sities and research laboratories with interests in geographic information topics.

[3] In fact, the Open Geospatial Consortium was before named Open GIS Consortium, but in September 2004 it was renamed in order to reflect better its current
aims.

metadata will have a direct influence in the performance of the services offered by these infrastructures.

This book is then focused in the technologies and methodologies that can provide a better utilization of metadata within spatial data infrastructures. In particular, this book will be centered on three main problems that hinder the correct utilization of metadata:

- The high volumes of geographic resources and the difficulty of cataloguing them correctly. Although many geographic resources have been created in last decades quite anarchically (and usually with no associated documentation), it is common to find that, at least, it is possible to identify group of related resources among these anarchical resources. There are collections or aggregation of geographic resources (or datasets) that can be considered as a unique entity from a general point of view. Most of these collections arise as a result of the fragmentation of geographic resources into datasets of manageable size and similar scale. The creation of metadata for this upper-level of collections palliates, in no small degree, the lack of documentation for the components of these collections. On the other hand, the hierarchical identification of collections and sub-collections (they can be organized in nested structures) facilitates the organization within a data repository. Imitating this physical organization of collections, catalogs should provide mechanisms to catalog collections of related resources, thus facilitating their navigation and creation of metadata.
- The diversity and heterogeneity of metadata standards. Along the last decade and as a response to the uncontrolled diffusion of geographic resources (and in general, all types of multimedia objects) encoded in disparate formats, many organizations (standardization bodies, software vendors, ...) started different initiatives for the definition of metadata standards to enable the common understanding within a community of users. However, despite the initial intention of common understanding, the diversity of initiatives originated also an undesired effect of heterogeneity. Nowadays, most of these initiatives have converged to a well defined international standard for each application domain. But despite this convergence there is still a need for facilitating interoperability between different metadata standards. On one hand, legacy metadata (the work done in the past) developed during years can not be directly thrown away. And on the other hand, visibility across different application domains is necessary to facilitate the reuse of resources. Spatial data infrastructures and Geolibraries (digital libraries specialized in geographic resources) are usually asked to provide a summary view (e.g., Dublin Core metadata) of their specific geographic metadata (e.g., ISO 19115), understandable by the general public or discovery agents.
- The heterogeneity of metadata content. By content heterogeneity it is meant the problem of identifying that the values given to a metadata element in two different metadata records are meaning the same concept

despite using different terms. When the metadata elements are constrained to a predefined list of values, there is no chance for heterogeneity. But if the domain (datatype) of a metadata element is free-text data, possible misunderstandings may appear. In fact, this problem is independent of the metadata schema used, i.e. we may have problems to identify that two metadata records are describing the same resource despite using the same schema. This situation implies that catalog discovery services can not be uniquely implemented as a simple word matching between the user queries and metadata records stored in the catalog. The idea is that discovery services should move from basic data retrieval strategies towards information retrieval strategies. Data retrieval consists mainly of determining which records in the catalog contain the words specified in the user query which, very frequently, is not enough to satisfy the user information need (Baeza-Yates and Ribeiro-Neto, 1999). On the opposite, information retrieval is concerned more with retrieving information about a subject than retrieving data which satisfies exactly a given query. Information retrieval systems usually deal with natural language text which is not always well structure and could be semantically ambiguous. Thus the integration of selected information retrieval techniques into metadata catalogs would help to understand the sense of the users vocabulary and to link these meanings to the underlying concepts expressed by metadata records.

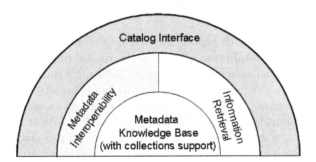

Fig. 0.1. Towards an enhanced catalog infrastructure

Therefore, the objective of this work will be to offer the proposals for incrementing the capacities of a metadata catalog infrastructure in three main aspects: the support of collections, the interoperability among different metadata standards, and the incorporation of information retrieval techniques. As depicted in figure 0.1, under a catalog interface layer we will propose:

- A solution for the management of nested collections. A Metadata Knowledge Base will be used as the basis of the catalog system infrastructure. The main features of this knowledge base are that it will support different metadata standards, and overall, that it will facilitate the management of collections of related resources. The metadata records describing the items of a collection are very similar. This work will investigate how to model and make profit of the aggregation relations that may be established among the metadata records describing the items and the entire collection. The hypothesis is that an appropriate modeling of these aggregation relations will enable the inference of meta-information, avoiding redundancies of information, and discovering new ways of browsing and monitoring collections of resources.

- A process for the construction of crosswalks between metadata standards. Crosswalks can be defined as the mechanisms or systems that enable the transformation between metadata in conformance with a source standard and the corresponding metadata in conformance with a target standard. Thanks to crosswalks, it will be possible to develop discovery services that search effectively across heterogeneous metadata holdings, i.e. they enable metadata interoperability.

- The use of selected vocabularies (disambiguated thesauri) and information retrieval techniques in order to improve the performance of catalog discovery services. This work will present a heuristic method for the semantic disambiguation of thesauri that are later used to fill the content of some metadata elements. These disambiguated thesauri will be used for the sense-based indexing of metadata records, thus enabling the application of classic information retrieval methods for the implementation of discovery services.

The three main research aspects covered by this book could be also managed from a more general point of view in the digital libraries context. However, the use of spatial data infrastructures as the application context increases the complexity of the proposals done in this book because of the inherent complexity of geographic information management (e.g., the management of geographic features, the complexity of geographic metadata standards, or the lack of appropriate geographic metadata corpora). On the positive side, this context enables an immediate application of the results to an industrial environment providing a very important feed-back for validating the research work done. Thus, this book results interesting not only for researchers and professionals in the Geographic Information domain, but also for those people involved in the more general domain of digital libraries.

Apart from this preface, this book is organized in other six chapters. The content of these chapters is the following:

- Chapter 1 presents the main issues related with the development of a spatial data infrastructure, making special emphasis in the role played by metadata.
- Chapter 2 is devoted to the solution proposed for the management of nested collections in catalog systems.

- Chapter 3 deals with the issues involved with the interoperability of metadata and presents a process to develop crosswalks to transform metadata.
- Chapter 4 remarks the benefits of using selected vocabularies (enhanced thesauri and ontologies) to fill the content of metadata elements in order to improve the performance of information retrieval in metadata catalogs.
- Chapter 5 presents the applicability of the previous concepts for the construction of components fully integrated within a spatial data infrastructure.
- And the final chapter 6 contains the summary conclusions and future research lines.

Zaragoza,
November 2004

Javier Nogueras-Iso
F. Javier Zarazaga-Soria
Pedro R. Muro-Medrano

Acknowledgements

There are many people to whom we are grateful for their support during the evolution of this book, many more than the ones we have space to acknowledge here.

First of all, we would like to thank to the members and friends of the Advanced Information Systems Laboratory (IAAA) of the University of Zaragoza during these years of hard work, making a especial mention to Pedro Álvarez, Rubén Béjar, Miguel A. Latre, Joaquín Ezpeleta, Silvia Laiglesia, David Portolés, Rodolfo Rioja, Javier Lacasta, David Anaya, David Gayán, Oscar Cantán, Rafael Tolosana, Covadonga Fernández, Pablo Gallardo, Rafael Martínez, Rubén Moreno, Jesús Barrera, Raquel Miguel and Javier López. All of them have contributed in some way (ideas, developments, tools, discussions, ...) to the successful end of this book. Additionally, we are also indebted to the other members of the TeIDE Consortium from the Jaume I University of Castellón and the Polytechnic University of Madrid because of their help and enthusiasm to boost the interest in Spatial Data Infrastructures in Spain.

We also want to remember here the people from the Geomatics & Remote Sensing General Office of the National Geographic Institute of Spain and especially to Sebastián Más and Antonio Rodríguez because of their interest and support.

We are very grateful to Michael Gould, Lars Bernard and José Angel Bañares, reviewers of some sections and chapters, and especially to Eva Méndez who made an extensive, deep, interdisciplinary and optimistic analysis. They found time in their busy schedules to read the text and provide valuable suggestions for its improvement.

Finally, we are absolutely grateful to our families and friends for all their patience, support and love. Much time have been stolen from our personal lives for the creation of this text. Undoubtedly, without their generous understanding this work would never have come into existence.

Contents

1   **Spatial Data Infrastructures and related concepts**   1
   1.1   Introduction   1
   1.2   Components of a Spatial Data Infrastructure................   3
   1.3   Integrating Digital Libraries concepts within Spatial Data
         Infrastructures   5
         1.3.1   Digital Libraries and Geolibraries....................   5
         1.3.2   Digital Libraries versus Spatial Data Infrastructures ...   8
   1.4   Metadata types and standardization initiatives   9
         1.4.1   Introduction   9
         1.4.2   General purpose metadata standards.................   10
         1.4.3   Metadata schemas for geographic resources   13
         1.4.4   Metadata schemas for service description   17
   1.5   Technical components of Spatial Data Infrastructures and
         the role of metadata   18
   1.6   Ontologies and Knowledge Representation in the context of
         Spatial Data Infrastructures   23
   1.7   Conclusions..   29

2   **A metadata infrastructure for the management of nested**
   collections   31
   2.1   Introduction   31
   2.2   Related work   35
         2.2.1   Addressing collections and relations in metadata
              standards   35
         2.2.2   Collections in Digital libraries and Geolibraries........   43
         2.2.3   Addressing relations in knowledge representation   47
   2.3   Defining the desired functionality of a collection enabled
         catalog system   49
   2.4   The Metadata Knowledge Base...........................   55
         2.4.1   Building the catalog services over a metadata
              knowledge base   55

        2.4.2 The knowledge base model 57
        2.4.3 Automatic generation of metadata................... 65
        2.4.4 Intelligent query answering 72
    2.5 Building aggregation relations........................... 82
    2.6 Conclusions and future work 85

3 **Interoperability between metadata standards** 89
    3.1 Introduction 89
    3.2 Related work 93
        3.2.1 Ontology based semantic interoperability 94
        3.2.2 Crosswalk based semantic interoperability 96
    3.3 Construction of crosswalks between metadata standards 99
        3.3.1 Harmonization 100
        3.3.2 Semantic mapping............................... 102
        3.3.3 Additional rules for metadata conversion 103
        3.3.4 Implementation of crosswalks: the use of style sheets ... 106
    3.4 Putting the method to work 115
        3.4.1 Transformation between CSDGM and ISO 19115 116
        3.4.2 Transformation between ISO 19115 and Dublin Core ... 120
    3.5 Conclusions and future work 125

4 **The use of disambiguated thesauri to improve information
    retrieval** 129
    4.1 Introduction 129
    4.2 Basic concepts about thesaurus and WordNet 133
        4.2.1 Thesaurus 133
        4.2.2 WordNet 135
    4.3 The Semantic Disambiguation of Thesauri 136
        4.3.1 State of the art in Semantic Disambiguation 136
        4.3.2 Description of the semantic disambiguation method.... 142
        4.3.3 Testing the method 148
    4.4 The information retrieval model 152
        4.4.1 State of the art in sense based information retrieval.... 152
        4.4.2 Introduction to the vector-space retrieval model 154
        4.4.3 The indexing of metadata records 155
        4.4.4 The indexing of queries 157
        4.4.5 Testing the retrieval model 158
    4.5 Conclusions and future work 166

5 **Integrating the concepts within the components of a
    Spatial Data Infrastructure** 169
    5.1 Introduction 169
    5.2 The catalog services component 171
        5.2.1 Introduction 171

5.2.2   Integrating the interoperability between metadata
        standards 173
5.2.3   Integrating the concept based information retrieval 176
5.3   A metadata editor 178
      5.3.1   Introduction 178
      5.3.2   Import/Export of metadata 181
      5.3.3   Collection Metadata Edition 181
      5.3.4   Thesaurus Management 185
5.4   The Web Portal of a Spatial Data Infrastructure 192
5.5   Conclusions and future work 194

6   Conclusions and future work 197

A   Collections 207
A.1   Consistency of the metadata model 207
A.2   Metadata Inference 217
      A.2.1   Generation of complete values...................... 217
      A.2.2   Update of whole-part hierarchy 219
      A.2.3   Example of a *wholeInferredValues* specification 221

B   Crosswalks 225
B.1   CSDGM→ISO 19115 stylesheet 225
B.2   ISO 19115→DC stylesheet 227
B.3   DC→ISO 19115 stylesheet 228

C   Applications 231
C.1   Revision of geographic metadata editors 231
C.2   Revision of thesaurus tools 236

References 241

Index 259

1

Spatial Data Infrastructures and related concepts

1.1 Introduction

This chapter is devoted to explain the main issues involved in the development of spatial data infrastructures. As a starting point, it must be mentioned that the origins of spatial data infrastructures can be found in the expansion of geographic information systems into a distributed and cooperative environment, not only from a technical perspective (advances in network technologies) but also taking into account cooperation policies among different organizations (public or private), and at different levels (local, regional, national or global).

Geographic Information System (GIS) is the term that is commonly used to refer to the software packages that are capable of integrating spatial and non-spatial data to yield the spatial information that is used for decision making. This includes computer-based equipment, procedures and techniques for manipulating spatial or map data. In this context, GISs are mostly used on a project basis, for example, to perform a particular analysis. When used in such a way, digital spatial data would be acquired by assembling the relevant maps and then digitizing or scanning them. And prior to the analysis, other data may be collected by using field techniques that collect the data in digital form. At this level, a geographic information system is used as a tool.

But data that are collected for a particular project are, in most cases, useful for other projects. This fact is even more pertinent with the recent "commoditization" of data and information. The costs involved with data collection are taken into account in project planning, along with attempting to maximize the use of the data from a project. Furthermore, it should be also realized that some data required for particular decisions are transient and may not longer be able to be collected when required. An example of this occurs when decisions concerning agricultural practices must be made. These decisions will often require environmental data spanning over several years. These data must be collected when they are available, even if the need for them is not present at the time of collection. Otherwise it is not possible to collect the data for past years when they are later needed. Thus there is a need to store this type of data in databases and make them accessible to others. These databases (spatial databases) become a shared resource, which must be maintained continuously. Moreover, the database, which has been maintained and exploited by a GIS tool, is itself often referred to as a geographic information system. Thus, at this level the own geographic information system may be viewed as a resource whose maintenance usually requires the cooperation and collaboration of several disciplines and a proper strategic plan. Furthermore, one might be interested in the interoperation of those resources (GISs), which are maintained at the state or national level, and sometimes by private corporations. In such cases, coordinating authorities are needed to assign custodianship and usage privileges for subsets of the data to different users (which may be agencies). Users in the general community are then able to expect the data to be available, and with network technology, to be accessible transparently. At this point, the geographic information systems have acquired the status of an infrastructure: a spatial data infrastructure.

The first formal definition of the term "National Spatial Data Infrastructure" was formulated in the US and published in the Federal Register on April 13, 1994 (U.S. Federal Register, 1994). It states: "National Spatial Data Infrastructure (NSDI) means the technology, policies, standards, and human resources necessary to acquire, process, store, distribute, and improve the utilization of geospatial data". The definition of Global Spatial Data Infrastructure (GSDI)[1] follows this closely. It states: "A coordinated approach to technology, policies, standards, and human resources necessary for the effective acquisition, management, storage, distribution, and improved the utilization of geo-spatial data in the development of the global community". Yet another view of a spatial data infrastructure is that of a system where the general community can expect the geo-spatial data to be available and accessible transparently with networking technology. In this view co-operation and collaboration between several disciplines together with the existence of a strategic plan for the maintenance of databases (which include spatial databases) are key components of the spatial data infrastructure.

The rest of the chapter is structured as follows. Section 1.2 presents an overview of the components that form part of a spatial data infrastructure. Although the political aspects are beyond of this scope of this book, this section stresses that cooperation policies and institutional arrangements have a direct influence in the development of spatial data infrastructures (Zarazaga-Soria et al., 2004; Nogueras-Iso et al., 2004b). Already from the technical point of view, section 1.3 indicates how the experience acquired in digital libraries may facilitate the development of spatial data infrastructures, especially in technological aspects (Zarazaga et al., 2000a). It must be remarked that spatial data infrastructures integrate multiple components which come from a background of different disciplines. Although spatial data infrastructures have been traditionally built over the concepts provided by geographic information systems,

---

[1] http://www.gsdi.org

they can be improved by means of the experience gained in other disciplines. In particular, spatial data infrastructures have some aspects (not all) in common with digital libraries specialized in geographic resources and services. Then, section 1.4 offers an overview of the different types of metadata and the standardization initiatives in this area. As it was mentioned in the introduction chapter, one of the essential pieces for the development of spatial data infrastructures (and the development of digital libraries in a wider context) is the use of metadata. Section 1.5 describes the technical components of a spatial data infrastructure and the role played by metadata in these components. Section 1.6 introduces the use of ontologies and knowledge representation in the context of spatial data infrastructures. Ontologies provide a shared and common understanding of a domain that can be communicated across people and application systems. Therefore, they are a popular research topic in this multidisciplinary environment of spatial data infrastructures, which aims at promoting the interoperability of data and services. And finally, this chapter ends with a conclusions section.

1.2 Components of a Spatial Data Infrastructure

According to (Coleman and Nebert, 1998), the main components of a spatial data infrastructure should include data providers (sources of spatial data), databases and metadata, data networks, technologies (dealing with data collection, management, search and representation), institutional arrangements, policies and standards, and end-users (see figure 1.1).

Let us see some details of about these components:

- Technology. Spatial data infrastructures should be developed over technological components created from the experience acquired working with generic information technology. One of the most important challenges should be the integration of all this experience, especially the one provided by the geographic information systems.
- Policies and Standards. Standards constitute the link among the different components of a spatial data infrastructures providing common languages and concepts that make possible their communication and coordination. Additionally, it is necessary the establishment of general guidelines to be followed by all the actors of a spatial data infrastructure. These guidelines should include several aspects such as architectures, processes, methods or standards.
- Human Resources. The development of spatial data infrastructures have to de done over the necessity of the users, both end-users and data providers (sources). On the other hand, the work to implement and maintain a spatial data infrastructure should be done by qualified teams of researchers and developers. All these people integrate the human resources that are necessary for the development of spatial data infrastructures.

- Institutional Arrangements. It is necessary the establishment of political decisions such as the creation of institutional framework. Agreements must be ratified to establish a national spatial data infrastructure, for coordinating the creation of regional spatial data infrastructures and for linking them to form the global spatial data infrastructure.
- Spatial Databases and Metadata. Spatial data infrastructures should be created over the geographic data, stored in the spatial databases, and their description (metadata).
- Data Networks. Spatial data infrastructures should be open systems deployed over data networks that provide the channel for accessing the services from remote systems.

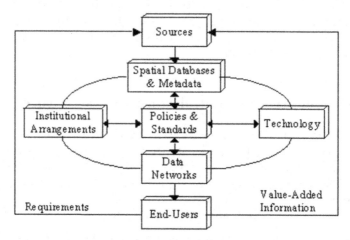

Fig. 1.1. A system view of the spatial data infrastructure components (taken from (Coleman and Nebert, 1998))

There are significant benefits managing the data-management problem from the spatial data infrastructure point of view. Firstly, it avoids the duplication of effort by ensuring all the stakeholders in the spatial data infrastructure are aware of the existence of data sets. Data providers are able to advertise and promote the availability of their data and potentially link to online services (e.g., services offering text reports, images, web mapping and e-commerce) that relate to their specific data sets. This way, all types of users (GI professionals or casual users) can locate all available geospatial and associated data relevant to an area of interest. On the other hand, the description of geospatial data with appropriate metadata builds upon and enhances the data management procedures of the geospatial community. Metadata helps organize and maintain the investment in data done by the entities participating in the spatial data infrastructure. Furthermore, reporting of descriptive metadata promotes the availability of geospatial data beyond the traditional geospatial community.

It must be remarked that spatial data infrastructures are just like other forms of better known infrastructures, such as roads, power lines or railways. The whole concept of spatial data infrastructures, and other forms of infrastructure, is that they allow authorized and/or participating members of the community to use them. They are simply available and taken for granted, although we may pay for the right to use them, for example through vehicle registration, railway tickets etc. Users essentially do not care how they work or who makes them work. In fact, it is said that the new spatial data infrastructures are being developed along similar lines as previous major transportation systems. Instead of transporting products and people by trains, planes or automobiles, digital networks transport ideas and information. The development of concrete infrastructures for the transport of things took decades, and continues today. The planning process was long and arduous. We must take a similar long view of the digital infrastructures of today, or we may see a breakdown similar to crumbling highways and broken water mains.

A spatial data infrastructure is the integration of multiple components which do not initially fit together in a seamless fashion for a number of reasons. Firstly, the necessary components come from a background of different communities and secondly, they should (in combination with other components) enable new functions which were not under consideration when the individual single components were designed and implemented. This means that the realization of large-scale globally spatial data infrastructures depends as much on collaborative effort as it does on the development of new technologies in order to develop systems which truly integrate their components. The level of collaboration required, across disciplines as well as across geographical boundaries, will be much higher than one could have previously envisioned.

As a conclusion, the new challenges related with the development of spatial data infrastructures should not be built over nothing. There is a very interesting background in several disciplines that can be used as a starting point for the creation of the new spatial data infrastructure concepts and methods. In particular, the experience acquired in digital libraries may facilitate the development of spatial data Infrastructures. In some aspects, a spatial data infrastructure can be considered as a digital library specialized in geographic resources and services. The following section studies this relation between digital libraries and spatial data infrastructures.

1.3 Integrating Digital Libraries concepts within Spatial Data Infrastructures

1.3.1 Digital Libraries and Geolibraries

There is not a uniform consensus about the definition and scope of digital libraries. In a traditional way, as it is mentioned in (Wiederhold, 1995), a digital library is popularly viewed as an electronic version of a public library, but replacing paper by electronic storage leads to three major differences: storage in digital form, direct communication to obtain material, and copying from a master version. From the information point of view, a digital library can be seen as a distributed text-based information system (Croft, 1995)(1), or as a distributed space of interlinked information (Schatz, 1995)(2). This view can be extended by defining a digital library as a large collection of electronic documents and the services that enable their use (Anderson, 1997)(3). In this sense, (Graham, 1998) provides a detailed definition of a digital library: "an organized data base of digital information objects in varying formats maintained to provide immediate ease of access to a user community, with these further characteristics: an overall access tool (e.g. a catalog) provides search and retrieval capability over the entire data base; and organized technical procedures exist through which the library management adds objects to the data base and removes them according to a coherent and accessible collections policy". As opposite to the largely unstructured information available on the Web, information in digital libraries is explicitly organized, described, and managed. In order to facilitate discovery and access, digital libraries systems summarize the content of their data resources into small descriptions, usually called metadata, which can be either introduced manually or automatically generated (index terms automatically extracted from a collection of documents). Additionally, some research works include in a digital library data, services and consumers who operate with them (Wilensky, 1995; Bertino et al., 2001)(4). This approximation involves the idea of a digital library like a set of electronic resources and associated technical capabilities for creating, searching, and using information; a community of users that construct, collect and organize the information; and functional capabilities for supporting the information needs and uses of that community.

It is possible to define a digital library specialized in geographic information resources. Usually, this kind of digital libraries is called Geolibrary. (Goodchild, 1998) defines the idea of "geolibrary" as a "library filled with georeferenced information" which is based upon the notion that information can have a geographic 'footprint'. Goodchild also explains that the geographic information system community has being working with geographic information, while georeferenced information is broader in scope to include such things as photographs, videos, music and literature that can be given a locational variable which defines a footprint. In this way, the idea of the geolibrary immediately extends well beyond the traditional scope of map libraries and archives to include almost all information contained with libraries; he later mentions that it can include information outside of libraries as well. This approximation has been used in projects like the Alexandria Digital Earth ProtoType (Janée and Frew, 2002). If we review the definitions presented before, a geolibrary could be understood as a distributed geographic-based information system (extending (1)). It is also a distributed space of interlinked information (2) because the data provided are related over geographic and thematic concepts.

Digital data can be seen as electronic documents and geolibraries provide services that enable their use (3). The digital information objects mentioned by Graham would be the geographic information resources. Finally, it is logical to assume that geolibraries include data, services and consumers who operate with them (4). However, there are some approximations indicating that traditional digital libraries do not have enough capacity for developing on-line information systems based on geographic information. For instance, (Gardels, 1997) suggests the incorporation of Open Geospatial Consortium propositions to digital libraries in order to complement them inside an environmental context. These deficiencies are directly related with the special complexity of geographic information. Some of these aspects could be:

- Geographic information has a very important visual component so the visualization aspects are one of the main keys of a geolibraries. In order to evaluate the data located it is necessary to see them. This visual aspect is not so relevant in classical digital libraries.
- Geolibraries should have the capacity for managing a large set of heterogeneous and complex objects. Every resource with some kind of geographic reference is a candidate object to be managed by a geolibrary. Apart from the classical geographic datasets, a geolibrary may store the publicity of a congress, books or photographs. Additionally, these data could be organized in complex structures with n-dimensions (e.g., spatial collections, temporal collections, etc.).
- The complexity of geographic metadata standards. Whereas a typical digital library metadata-standard like Dublin Core (DCMI, 2004; ISO, 2003d; ANSI, 2001) has only 15 elements, the typical geographic information metadata-standard ISO 19115 (ISO, 2003a) has more than 350 elements. Additionally, this is not only a scale-complexity problem, but it also implies a fine-grained description of resources. For instance, Dublin Core only defines one item, the *subject* element, for describing the topics of the data content. However, ISO 19115 provides six items to include the same meta-information: a *topic category* element and five types (*discipline, theme, place, stratum* and *temporal*) of *descriptive keywords*. This has implications from the information-discovery point of view because the search systems that operate in a geolibrary should have the skill for managing this complexity transparently to the end-user.
- In most cases, geographic information resources require some kind of pre-processing before being delivered to the user. Frequently, the resource downloaded from the spatial data infrastructure is incorporated to a geographic information system project and it must comply with the same constraints as the rest of layers (coverages) that form part of the project. For instance, the spatial data infrastructure should be able to perform automatically coordinate reference systems transformations (projections, datums, ...) on user demand. This would imply the automatic modification of the piece of metadata describing the transformed features as well.

1.3.2 Digital Libraries versus Spatial Data Infrastructures

The next question that it is necessary to make is "are geolibraries and spatial data infrastructures the same thing?". (Boxall, 2002) indicates that it results curious that the released CookBook for GSDI (Nebert, 2001) makes almost no mention of libraries, and even any mention is quaint. He argues that the reason could be the lack of involvement by librarians in GSDI. Nevertheless, when this document deals with technical problems, references to libraries (digital/electronic or not) are presented. Geolibrary experiences could provide knowledge for the development of concepts in many of the main components of a spatial data infrastructure mentioned before. In this sense, geolibraries could be considered as a part of the spatial data infrastructures (especially as a technological basis).

However, spatial data infrastructures most reinforce some aspects of geolibraries to make this technology and knowledge suitable for the more demanding context of spatial data infrastructures. Some of these special characteristics could be:

- Spatial data infrastructures have an important political and social component. As a consequence of this, public administrations at different levels are usually involved in their creation and maintenance. As a result of developing spatial data infrastructures at different levels, a model of spatial data infrastructure hierarchy that includes spatial data infrastructures developed at different political-administrative levels was developed and introduced by (Rajabifard et al., 2000). This model presents a spatial data infrastructure hierarchy is made up of inter-connected spatial data infrastructures at corporate, local, state or provincial, national, regional and global levels. In the model, a corporate geographic information system is deemed to be a spatial data infrastructure at the corporate level, the base level of the hierarchy. Each spatial data infrastructure at the local level or above is primarily formed by the integration of spatial datasets originally developed for use in corporations operating at that level and below.
- The standardization processes in spatial data infrastructures involve not only the organization of data, but also issues related to the capture and integration of these data. These problems are inherent to digital libraries, but are aggravated in the case of georeferenced data. Moreover, users of geographic applications have a wide range of requirements for visualization and manipulation of the data. For instance, given the heterogeneity of users (e.g., biologists, ecologists, architects, engineers, demographers), both the vocabulary used to search for data and the presentation format for the selected data are specific for each user profile and application.
- Spatial data infrastructures are much more concerned with data maintenance. Geographic information resources are updated continually due to typical causes like error-corrections or evolution of data (e.g. modifications to reflect new ground conditions). These frequent modifications involve the

necessity of the maintenance of the corresponding metadata, and the catalogs which store them. Additionally, geographic information is often used as a source for creating new geographic information. In this way, metadata may be used for deriving new metadata.

- In general, digital libraries act as local repositories of the digital resources that are published in the library. In contrast, the gateways or catalogs provided by a spatial data infrastructure work in many cases as brokers for third party organizations or spatial data infrastructures at a narrower level. The spatial data infrastructure has to facilitate mechanisms for accessing transparently the information which may be stored either locally (the information producer and the spatial data infrastructure manager are the same entity), in a remote repository, or distributed by third party organizations. This is an important and still little explored research area.
- Spatial data infrastructures not only include the delivering of the digital resource, but they also provide services that exploit these data (gazetteer, web map visualization, service chaining, etc.). On the contrary, digital libraries are usually only concerned with the delivering of the digital resource as raw data.
- Integration from and to legacy systems. Geographic information is used in many situations to create new geographic information. In this case, legacy systems used for creating this kind of data should be provided by utilities to incorporate their results to the spatial data infrastructure (metadata creation, information catalog, etc.). On the other hand, geographic information can be used as a base of decision systems such as alert management systems, geo-marketing, etc. In this case, legacy systems used for decision making should be provided with tools for locating and using the "best" information available through the spatial data infrastructure.
- Necessity of interoperability with other digital library systems. As a consequence of the necessity for managing heterogeneous information, spatial data infrastructure services could be developed over the services provided by other digital library systems.

1.4 Metadata types and standardization initiatives

1.4.1 Introduction

Most commonly defined as "structured data about data" or "data which describes attributes of a resource" or, more simply, "information about data", the concept of metadata is not new: map legends, library catalog cards and business cards are everyday examples. Basically, metadata offers description of the content, quality, condition, authorship, and any other characteristics of the resources. It also provides for a standardized representation of information. That is, similar to a bibliographical record or map legend, it provides a common set of terminology to define the resource or data. Metadata constitute the mechanism to characterize data and services in order to enable other users or applications to make use of such data and services. Metadata records, each one describing a specific resource, are grouped into catalogs thus providing the users with the possibility of finding the resources of their interest. Therefore, these catalogs are the tool to put in touch consumers with information producers.

One circumstance that must be taken into account is that the structure of metadata descriptions varies according to the type of resource. Therefore, recognized organizations in each application domain establish a specific structure of metadata, also called metadata schema or metadata standard. For instance, MARC(MAchine-Readable Cataloguing) is one of the most widely used standards in the library application domain. It defines a data format which emerged from a Library of Congress led initiative begun thirty years ago. MARC became USMARC in the 1980s and MARC 21 in the late 1990s. It provides the mechanism by which computers exchange, use and interpret bibliographic information and its data elements make up the foundation of most library catalogs used today (U.S. Library of Congress, 2004b).

Next subsections will detail different metadata schemas used in the geographic information application domain: schemas for the description of geographic resources; schemas for the description of services; and other general purpose standards that although do not directly describe geographic information, are being also used in this context.

Finally, apart from the chosen metadata-standard, it must be mentioned that metadata cataloguing systems usually support (recognize) three forms of metadata (Nebert, 2001): the implementation form (within a database or storage system), the export or encoding format (a machine-readable form designed for transfer of metadata between computers), and the presentation form (a format suitable to viewing by humans). For the last two forms, there is a general consensus about the use of XML (eXtensible Markup Language) (Bray et al., 2000). First of all, it includes a capable markup language with structural rules enforced through a control file, in the form of in the form of a DTD (Document Type Definition) or an XML-Schema (an enhanced version of DTD) (Thompson et al., 2001). Organizations in charge of the edition of metadata standards publish stable versions of DTDs and XML-Schemas in order to assure the conformance of metadata descriptions in XML format. And secondly, through a companion specification (XML Style Language, or XSL) (W3C, 2004a), an XML document may be used along with a style sheet to produce flexible presentations or reports of content according to user requirements.

1.4.2 General purpose metadata standards

A good example of a simple general purpose metadata standard is the one proposed by Dublin Core Metadata Initiative (DCMI) (DCMI, 2004). This initiative, created in 1995, promotes the widespread adoption of interoperable metadata standards and the development of specialized metadata vocabularies that enable more intelligent information discovery systems. The Dublin Core metadata element set is a standard for the description of cross-domain information resources, i.e. any kind of resource, regardless of the media format, area of specialization or cultural origin. This set consists of 15 basic descriptors which are the result of an international and interdisciplinary consensus. Nowadays, the Dublin Core metadata element set has become an important part of the emerging infrastructure of the Internet. Many communities are eager to adopt a common core of semantics for resource description, and the Dublin Core has attracted broad ranging international and interdisciplinary support for this purpose. The Dublin Core now exists in over 20 translations, has been adopted by CEN/ISSS (European Committee for Standardization / Information Society Standardization System), and is documented in two Internet RFCs (Requests For Comments). It has also official standing within the WWW Consortium and the Z39.50 standard. Dublin Core metadata has been approved as a U.S. National Standard (ANSI/NISO Z39.85) (ANSI, 2001), formally endorsed by over seven governments for promoting discovery of government information in electronic form, and adopted by a number of supranational agencies such as the World Health Organization (WHO). Numerous community-specific metadata initiatives in library, archival, educational, and governmental applications are using the Dublin Core as their basis. Moreover, since April 2003, the Dublin Core Metadata Element Set standard has been adopted as ISO standard (ISO 15836) (ISO, 2003d). This approval is the culmination of an incremental process to bring the Dublin Core metadata element set into a worldwide audience. As an international standard, it will be easier for many organizations to adopt and promote the use of Dublin Core to enhance resource discovery on the Internet.

On the other hand, given the simplicity of Dublin Core metadata, its 15 metadata elements typically overlap 1:1 with any broader schema for metadata (see chapter 3 for further details about mappings among standards). Therefore, more and more organizations in the geographic information domain are considering the adoption of Dublin Core for some of the following uses: to serve as an interchange format between various systems using different metadata standards/formats; to use for harvesting metadata from data sources within and outside of the library domain; to support simple creation of library catalog records for resources within a variety of systems; to expose CSDGM or ISO 19115 data to other communities (through a conversion to DC); to allow for acquiring resource discovery metadata from non-geographic information creators using DC.

The document "DCMI Metadata Terms" [2] defines the current list of metadata elements, qualifiers and vocabulary terms. According to the inclusion or exclusion of some of these metadata terms, Dublin Core metadata are classi-

---

[2] http://dublincore.org/documents/dcmi-terms/

fied into two categories: Simple Dublin Core metadata and Qualified Dublin Core metadata. By Simple Dublin Core Metadata it is meant metadata records which contain uniquely elements that belong to the Dublin Core metadata element set, and which do not use element qualifiers. And by Qualified Dublin Core metadata it is meant the rest of metadata records that may use qualifiers. These qualifiers may be of two types: element refinements (these qualifiers make the meaning of an element narrower or more specific) or encoding schemes (schemes defining the possible values of an element which facilitate the element interpretation).

Anyway, although maintaining the simplicity of Dublin Core, it is also possible to define particular profiles of Dublin Core metadata for specific domains. The concept of application profiles has emerged within the Dublin Core Metadata Initiative as a way to declare which elements from which metadata schemas (uniquely identified by means of namespaces) are used in a particular application or project. For instance, a CEN/ISSS workshop has developed a geo-spatial application profile of Dublin Core (Zarazaga-Soria et al., 2003b). This workshop is called MMI-DC (Metadata Multimedia Information - Dublin Core) workshop and the author of this work has collaborated actively for the creation of this profile. This geospatial application profile is a specification that defines: the elements and refinements taken from the general Dublin Core model; the domain type of the element values (e.g., specifying the use of a specific controlled vocabulary or encoding scheme); the additional elements and qualifiers that are taken from other application profiles; the refinement of standard definitions; and the conditionality and occurrence of elements.

Additionally, it must be noticed that the document "DCMI Metadata Terms" is an abstract specification of metadata content but when metadata is exchanged, it is usually encoded either as HTML $< meta >$ tags (suitable for embedding in the $< head > ... < head >$ section of the page) or as part of RDF(Resource Description Framework) descriptions (Manola and Miller, 2004). The intended use of $< meta >$ tags was to describe the content of a Web page, thus making this meta-information visible to search engines. However, current search engines hardly trust, at least entirely, in this meta-information for the indexing of Web pages. On the contrary, the second possibility, RDF, is becoming increasingly important because it is one of the underlying technologies in the new conception of the Web: the Semantic Web. According to (W3C, 2004b; Berners-Lee et al., 2001), "the Semantic Web is an extension of the current web in which information is given well-defined meaning, better enabling computers and people to work in cooperation".

RDF is a W3C recommendation for modeling and exchanging metadata, which is expressed in XML format. The major advantage of RDF is its flexibility. RDF is not really a metadata standard defining a series of elements. On the contrary, it can be considered as a meta-model that contains other metadata schemas or combinations of them. RDF uniquely defines a simple model for describing the interrelationships among resources in terms of named properties and values. But for the declaration and interpretation of those properties, a complementary technology of RDF is needed. This complementary technology is RDFS, which stands for RDF Schema although it has been recently renamed as RDF Vocabulary Description Language (Brickley and Guha, 2004). RDFS provides a rich set of constructs to define and constrain the interpretation of vocabularies used in a certain information community. Thus RDFS provides the technology to define descriptive vocabularies like Dublin Core metadata.

1.4.3 Metadata schemas for geographic resources

Geographic metadata is the description of a particular geographic dataset. As it is referred in (Nebert, 2001): "Metadata helps people who use geospatial data find the data they need and determine how best to use it". Maybe one of the features that distinguish the geographic metadata with respect to other types of metadata is that its creation and maintenance is a hard and thorough process, which requires time and important human efforts. However, as stated in (FGDC, 2000), the creation of metadata has three major objectives that derive in three important benefits:

- The first one is to organize and maintain the investment in data made by an organization. As personnel change or time passes, later workers may have little understanding of the content and uses for the digital data previously created and may find that they cannot trust results generated from these data. That is the reason why complete metadata descriptions of the content and accuracy of a geospatial data set will encourage appropriate reuse of the data. It may seem burdensome to add the cost of generating metadata to the cost of data collection, but in the long run the value of the data is dependent on its documentation (Nebert, 2001). Moreover, such descriptions may also provide some protection for the producing organization if conflicts arise over the misuse of data.
- The second objective is to provide information to data catalogs and clearinghouses. Applications of geographic information systems often require many themes of data. However, few organizations can afford to create all data they need on their own. Often data created by an organization may be also useful to others and by making metadata available through data catalogs and clearinghouses, organizations can find: data to use; partners to share data collection and maintenance efforts; and customers for their data.
- And finally, the third objective of metadata is to provide information to aid data transfer. In fact, metadata should accompany the transfer of a data set. In this way, metadata aids the organization receiving the data process and interpret data, incorporate data into its holdings, and update internal catalogs describing its data holdings.

In order to extend the use and understanding of metadata through different communities of users, e.g. to enable distributed searches across a network catalog servers, it is necessary to use well-defined contents and thus adjust them to a metadata standard. In this way, there are several standard proposals to describe consistently a geographic resource, which have arisen at national or global level and with different scopes. Some of the most extended ones are:

- The Content Standard for Digital Geospatial Metadata(CSDGM) (FGDC, 1998, 2000). It was carried out in 1994 by the Federal Geographic Data Committee (FGDC) of the United States to give support for the construction of the National Spatial Data Infrastructure. And although it is a national standard, it is the oldest one and has been incorporated into many GIS tools and networks (e.g. the Clearinghouse project), thus becoming the most widely used in GIS world (e.g. adopted in countries like South Africa or Canada).
- The European voluntary norm prENV 12657 (CEN, 1998). The European Committee for Standardization also created a technical committee, the CEN/TC 287, for the establishment of Geographic Information European pre-standards. In 1998 this committee published a version of the 12657 norm which was adopted in several projects like GDDD (Geographic Data Description Directory), LaClef or ESMI (European Spatial Metadata Infrastructure), all of them encouraging the access to geographic resources [3].
- The international standard ISO 19115 (ISO, 2003a). The organization responsible for this standard is the International Organization for Standardization who created in 1992 the committee 211 (ISO/TC 211) with responsibilities in "geomatics" [4]. This committee is now preparing a family of standards that, in the near future, will obtain the rank as official international standard. One of these standards is the Nr. 19115, released as standard in May 2003, which defines the schema required for describing geographic information and services. It provides information about the identification, the extent, the quality, the spatial and temporal schema, spatial reference, and distribution of digital geographic data. This standard is applicable to: the cataloguing of datasets, clearinghouse activities, and the full description of datasets; geographic datasets, dataset series, and individual geographic features and feature properties. Furthermore, though ISO 19115 is applicable to digital data, its principles can be extended to many other forms of geographic data such as maps, charts, and textual documents as well as non-geographic data.

Apart from these main standards, there are other metadata standard initiatives arisen at a regional, national or domain-specific level like: the Spanish norm for geographic information exchange known as MIGRA (Exchange Mechanism for Relational Geographic Information constituted by Aggrega-

---

[3] More details of these projects can be found at http://www.eurogeographics.org/.
[4] Homepage of ISO/TC 211 (Geographic information/Geomatics): http://www.isotc211.org.

tion) (AENOR, 1998); the UDK metadata standard from the German Environmental Data Catalog[5]; or GELOS (Global Environmental Information
Locator Service) from the G-7/G-8 Environment and Natural Resources Management project [6].

The intention of the different organizations who have proposed these
schemas is the harmonization of all the initiatives around ISO 19115, which
has been recently released as standard. However, at this moment, the FGDC
standard (CSDGM) is the most widely used in GIS world and there exist
multiple tools for the creation of metadata. It is the oldest one and has been
popularized by the development of the North American Spatial Data Infrastructure and its National Geospatial Data Clearinghouse project[7]. On the
other hand, all of these standards share a common core of metadata elements
and it is possible to construct systems talking these different standards and
enabling interoperability. In order to give an idea of the different aspects of
data covered by this common core, the following sections could be mentioned:

- A metadata reference information section provides administrative information about the own metadata record: authorship of metadata record,
  creation and last update date, name and version of metadata standard or
  the language of metadata descriptions.
- An identification information section contains basic information such the
  title of the data set, references to the originators of data, what geographic
  area it covers, keywords, status of data or information about access and
  legal constraints. Most elements included in this section have an equivalent
  element in not specific geographic metadata standards like Dublin Core.
- A data quality section usually describes how good the data are, details
  positional and attribute accuracy, or explains the process and sources that
  originated the dataset.
- Sections like spatial data organization and spatial reference system compile
  the specific characteristics of pure spatial information, that is, the spatial
  data model that was used to encode the spatial data (vector, raster,) or
  other possible methods for indirect geo-referencing (street addresses, postal
  codes, etc.) and information about the spatial reference system (datum,
  ellipsoid, projection) used, e.g. geographic coordinates (longitude/latitude)
  or a UTM Zone 30 N projected coordinate system.
- An entity and attribute information section informs about the features
  (roads, houses, elevation, temperature, etc.) included in the datasets, their
  attributes and the encoding methods for the domain values (codes used
  and meaning,).
- Finally another basic section is distribution information, i.e. who dis-
  tributes the data, what formats are used, availability and price of data
  and so on.

---

[5] http://www.umweltdatenkatalog.de/

[6] http://www.g7.fed.us/enrm/

[7] http://www.fgdc.gov/clearinghouse/clearinghouse.html

With independence of the standard being used, it is also usual to classify the metadata elements according to their role of data access in an end-to-end resource discovery, evaluation and access paradigm (Nebert, 2001). Although metadata standards do not separate explicitly the elements according to these roles, three incremental levels of metadata could be distinguished:

- Discovery metadata is the minimum amount of information that needs to be provided to convey to the inquirer the nature and content of the data resource. This falls into broad categories to answer the "what, why, when, who, where and how" questions about geospatial data. Typical discovery metadata elements could be the title and description of the dataset or its geographical extent.
- Exploration metadata provides sufficient information to ascertain that data fit for a given purpose exists, to evaluate its properties or to reference some point of contact for more information.
- And finally, exploitation metadata includes those properties required to access, transfer, load, interpret, and apply the data in the end application where it is exploited.

Another important aspect concerning metadata schemas is the level of detail, which is defined by the election of the standard and the creation of special extensions and profiles. Firstly, the elected standard comprises a bigger or smaller set of elements with different conditionality: mandatory (and mandatory if applicable) elements which must be completed to deliver metadata entries compliant with the standard; and optional elements which allow for a more extensive standard description of geographic data, if required. And secondly, extensions and profiles of the standard may be defined. An extension of the standard typically consists of the additional constraints (e.g. certain optional elements that become mandatory, modify the repeatability of an element), special codes and the creation of new elements and entities. ISO 19115 and CSDGM provide methods within the own standard for extending metadata to fit specialized needs. And if these additional features are extensive, involving the creation of many metadata elements within a metadata entity, specific to a discipline or application, ISO recommends the creation of a community profile to coordinate the proposed extension via formalized proceedings user groups.

However, although specific profiles and the conditionality of elements enable certain flexibility of geographic metadata standards, they result still very detailed. CSDGM and ISO 19115 standards comprise more than 300 elements distributed in sections and subsections. The problem is that in order to complete metadata records in accordance with such detailed standards, metadata creators must be highly qualified and spend quite a lot of time. That is the reason why the own document defining ISO 19115 standard also includes a profile called "Core metadata for geographic datasets" that only includes 22 elements, the minimum number of metadata elements required to identify a dataset, typically for catalogue purposes. Other organizations involved in the cataloguing of geographic resources go even further and propose the use of more generic standards like Dublin Core (DCMI, 2004), a metadata standard of general outreach that enables a minimum description of resources (see section 1.4.2). Its 15 metadata elements typically overlap 1:1 with any broader scheme and it results more convenient to have just 15-20 metadata elements correctly completed and up-to-date, than 300 uncontrolled elements. The idea is to provide at least discovery level metadata.

1.4.4 Metadata schemas for service description

Nowadays, the development of the services offered by spatial data infrastructures, and in general the development of services in any type of networked infrastructures, is usually guided by the Web Services Architecture (Booth et al., 2004) proposed by W3C. This architecture aims at providing a standard means of interoperating between different software applications (the Web services), running on a variety of platforms and/or frameworks. According to the W3C Web Services Glossary (Haas and Brown, 2004), a Web service is a software system designed to support interoperable machine-to-machine interaction over a network. Furthermore, it is specifically mentioned that its interface must be described in a machine-processable format in order to enable other systems to interact with the Web service in a manner prescribed by its description using SOAP-messages (Box et al., 2000), typically conveyed using HTTP (Fielding et al., 1999) with an XML serialization in conjunction with other Web-related standards.

Therefore, the great impact of Web Services has increased the importance of metadata that describes the processing capabilities of services. The details of a Web Service can be published in a catalog, so that a client's (or another service's) request for such a service can lead to the client invoking that service.

The leading and most accepted standard for service metadata is WSDL (Web Services Description Language) (Christensen et al., 2001). WSDL is a means of describing a service connection (operation signature or binding) for software to connect to it. Service Directory specifications like UDDI (OASIS, 2004) can use WSDL to express the machine-readable connect to a service. The main disadvantage of WSDL is that it does not have the ability to characterize content very well. Another emerging proposal is DAML-S (Web Service Ontology)(Burstein et al., 2002), which extends DAML+OIL (van Harmelen et al., 2001) for describing web services (see section 1.6 for a more detailed description about ontology and DAML+OIL). The goal of DAML-S is to provide software agents with computer interpretable descriptions of web services in order to enable automatic discovery, invocation, composition, and execution monitoring of web services.

In the context of geographic information services, there is still no stable standard. In principle ISO 19115 standard (ISO, 2003a) was also intended for the description of services but it lacks details for the specific methods of these services. On the other hand, OGC has issued a draft specification for a Services

Catalog (Reich and Vretanos, 2001) that uses as metadata schema a proposal from ISO 19119 standard (ISO, 2002). This standard aims at providing a taxonomy of geographic information services and it provides an abstract level description of services with fields similar to those in ISO 19115. However, similar to ISO 19115 standard, this proposal results insufficient to detail the bindings of Web Services. In fact, last tendency of geographic information services metadata is to move towards more generic standards like WSDL, not specific to the geographic information application domain (Lieberman, 2003).

1.5 Technical components of Spatial Data Infrastructures and the role of metadata

As it has been mentioned in the introduction of this chapter, spatial data infrastructures are a solution to manage efficiently geographic information (Nebert, 2001). The main goal of this kind of infrastructures is to facilitate and enable an efficient exploitation of geographic information to the multiple stakeholders in geographic information market, either at global, national of local level. And for the development of the different components of such infrastructures, it is essential the management of metadata for description of geographic resources and services.

Figure 1.2 presents the components architecture of a typical spatial data infrastructure serving the data needs of an institution (e.g., an institution in charge of controlling the hydrographic resources of a river basin (Arqued-Esquía et al., 2001; Nogueras-Iso et al., 2004b)). In the figure it is also shown the operational view on an intranet/internet environment, it displays the relationships with other institutions SDIs (e.g., other authorities in charge of river basin resources or the environment ministry of a nation). As it can be observed, each institution controls its local SDI and in turn is a node belonging to a higher-level SDI. Apart from the special characteristics of every component depicted in the figure (which are later detailed), the most important feature of such infrastructures is that it should not be an ad-hoc system to solve a specific problem. Quite the opposite, the geographic information services are intended to be compatible and interoperable with services belonging to other SDI nodes, which are subject to be created by any kind of organization either at local, national or global level. Concerning this interoperability aim, it should be mentioned the existence of two main standardization initiatives that have arisen in last years: the ISO committee with responsibilities in geomatics (ISO/TC 211) and the Open Geospatial Consortium (OGC). From a more abstract perspective, ISO/TC 211 provides general purpose standards and specifications defining relevant aspects of the description and management of geographic information and geographic information services. And on the other hand, OGC develops and provides, through a membership submission and consensus process, implementation-level technical specifications for interfaces to geospatial processes and geospatial information, most of them based on emerging standards in ISO/TC 211. That is to say, both initiatives collaborate in the definition process of the standardized interfaces for the main services that are typically integrated within a spatial data infrastructure and with special focus on defining services accessible via Web.

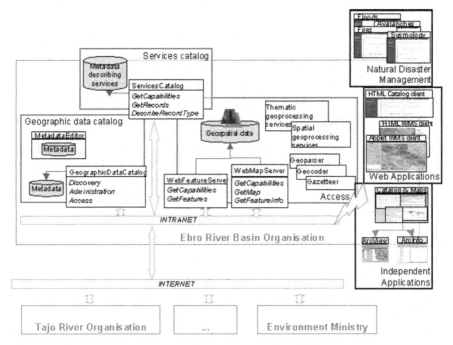

Fig. 1.2. Architecture of an SDI

The most important part of an SDI is the Geographic Data Catalog components area. Geographic data catalogs (Kottman, 1999) are the solution to publish descriptions of geospatial data holdings in a standard way that enables search across multiple servers. They enable users to locate the spatial data of their interest. These descriptions of resources are called metadata and are used by catalog discovery services as the target for query on raster, vector, and tabular geospatial information. Furthermore, the use of indexed and searchable metadata provides a selected and disciplined vocabulary against which intelligent geospatial queries can be performed, thus enabling the understanding among users from the same community or even belonging to different geographic information communities (section 1.4.3 revises the different metadata schemas used to describe geographic information). In summary, this area encompasses the necessary components that enable the SDI stakeholders to create metadata, and publish them thus facilitating the search to the intended audience.

One of the leading specifications proposed for catalog interoperation is the OGC Catalog Interface Implementation specification (Nebert, 2002; Muro-Medrano et al., 2003; Nogueras-Iso et al., 2004c), which describes the set of service interfaces that support Management, Discovery, and Access to geospatial information. Firstly, discovery services allow users to search within the catalog using a query language with a recognized syntax. The specification provides two main possibilities for the query language: the OGC Common Query Language (similar to the specification of WHERE clauses in SQL) or RPN based languages (queries represented in Reverse Polish Notation format). However, it does not exclude the use of other languages. In fact, last versions of other OGC specifications promote the use of the Filter Encoding Specification (Vretanos, 2001), which is based on XML language. Moreover, thanks to the advances in XML technologies, this language results much more suitable for catalog implementation or for the creation of user interfaces that facilitate query construction. Secondly, Management services provide functionality for the management and organization of metadata catalog entries maintained in a local storage device (e.g., file systems or Relational Database Management Systems). And thirdly, the role of Access services is to enable access to items which have been previously located through Discovery services. Access to geospatial data from the consumers point of view is a part of a process that goes from discovery to evaluation, then to access, and finally to exploitation. Finally, another important aspect in this specification is that it provides different profiles for these interfaces according to the distributed computing platform where they are going to be implemented. In particular, within this specification the profiles for CORBA (Orfali et al., 1999), WWW and OLEDB/COM (Gordon, 2000) are provided. In particular, the WWW profile is the most widely accepted because it is compatible with the search and retrieval protocol Z39.50 (ANSI, 1995), very popular in the world of digital libraries. This WWW profile, based on a message-passing client/server architecture, establishes a mapping between each one of the operations belonging to the catalog interface general model and the corresponding service specified by the norm ANSI/NISO Z39.50 (also known as ISO 23950). Z39.50 encodes protocol messages as byte streams over TCP (Basic Encoding Rules - BER (ISO, 1990)). However, this WWW profiles adds the possibility of encoding the messages in XML over HTTP (using XER [8]), thus avoiding firewall restrictions problems and following the current tendency of creating Web services over HTTP.

Another important area in the construction of an SDI is the Geospatial Data Access components. Once spatial data of interest have been located, it is necessary to visualize and evaluate the data. Then if this is the data desired, advanced users will require the geospatial data in its packaged form. This area integrates various components that conform to interoperable standards. The Web Map Server component provides mapped or graphical views of geospatial

---

[8] XML Encoding Rules (XER). Available at http://asf.gils.net/xer.

data through online mapping interfaces (Beaujardière, 2002; Fernández et al., 2000). This way, it is possible to evaluate data and satisfy many of the needs of the users without requiring the full data download. For the final access in its packaged form, this area also integrates a Web Feature Server (Vretanos, 2002) component, which provides geographic information in GML (Geographic Markup Language) (Cox et al., 2003). This area will also integrate components which provide services for geo-processing and that are currently in the process of standardization. Some examples of these services could be: spatial geo-processing for coordinate transformation, format conversion and combining of different geospatial resources; thematic processing for geoparameter calculation, thematic classification or subsetting and subsampling based on parameter values; temporal geo-processing to provide subsetting and subsampling based on temporal values; and association services like Gazetteers, Geocoders or Geoparsers.

Another component vital to interrelate spatial data infrastructures is the Services Catalog area. In order to make accessible the services offered by an SDI to the general public or to other SDIs, it is necessary to maintain a record of available services. The Services Catalog component offers this directory service that publishes the distinct services (Web Map Servers, Web Feature Servers, Geographic Data catalogs ...) within a distributed network. For instance, this component can monitors the Geographic Data Catalogs connected to the network and serve as a portal which redirects queries to the different nodes in the network. There is still no established specification for a services catalog. Until 2001, OGC published several specification drafts describing a set of interfaces similar to those provided for a geographic data catalog. These draft specifications were stateless protocols intended for light clients and servers supporting HTTP 1.0 that used metadata descriptions of services as query target. Regarding the metadata schema used, these specifications used a proposal from ISO 19119 standard (ISO, 2002) which aims at providing a taxonomy of geographic information services. And regarding the query language, they also allowed the use of OGC Common Query Language, RPN based query languages, and Filter Encoding Specification.

However, since beginning of 2002, OGC has changed radically the strategy for services catalogs. On one hand, OGC tends to unify data catalogs and services catalogs. The OGC refers to these catalogs as "geospatial catalogs" because they describe or refer to geospatial content and/or services. According to the last draft of the OGC Catalog specification (Nebert and Whiteside, 2004), a geographic data catalog can also be extended to store and maintain service metadata. And on the other hand, instead of the design of ad-hoc HTTP message-passing protocols, it is believed that future versions of OGC Catalog Specifications will establish a methodology for "stateless" catalog transactions following a true Web Services architecture (Graham et al., 2002). In (Lieberman, 2003) OGC recommends that new versions of OGC services should comply with underlying web technologies such as: SOAP (Simple Object Access Protocol) (Box et al., 2000) for RPC (Remote Procedure Call) communication between client and servers; WSDL (Christensen et al., 2001) to describe service capabilities; or UDDI (Universal Discovery, Description and Integration) (OASIS, 2004) as the standard for service registries.

Finally, another important issue in the development of an SDI is the one concerned with the clients of the infrastructure. The right part of figure 1.2 shows three types of clients that make profit of the services offered by the infrastructure. The first group of clients integrates a generic search client (applets or HTML light-weight clients) as well as clients of Web Map Servers to visualize the data discovered by means of the Catalog. Another group of clients offers a customization of the generic search tools for a specific context, e.g. Natural Disasters Management. And thirdly, the clients entitled as independent applications represent other applications that combine access to the catalog infrastructure and the integration with commercial GIS tools (e.g. ArcView, ArcInfo from ESRI) (Nogueras et al., 2000; Cantán et al., 2001b,a).

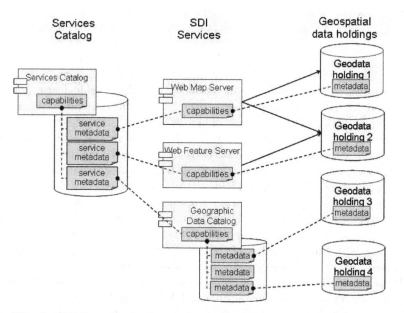

Fig. 1.3. Connection between metadata and the different SDI services

One conclusion that can be extracted from the aforementioned services is that all of them are based at a higher or lower scale on the use of metadata, which is either used to describe the own service functionality or used as data managed by this service. Figure 1.3 shows another perspective of geographic information services which is focused on the data and metadata used by each service. The right side part of the figure shows the geospatial data holdings (created by GI providers) together with their related metadata (ideally completed during geodata creation process). In the middle, the services which either publish geospatial data descriptions or provide access to data are depicted. A geographic data catalog is an example of a service that manages or publishes directly geospatial data descriptions, i.e. metadata descriptions of geospatial data holdings. An on the other hand, Web Map Servers or Web Feature Servers are examples of services providing on-line access to geospatial data. Nevertheless, these services also manage a special type of metadata called capabilities. These capabilities describe the functionality of the service together with a short summary of the different data layers offered by the service, which can be derived from the metadata descriptions on the right. Finally, on the left of the figure the Services Catalog component serves as a directory (registry) of the services offered by an SDI node. Once again, the metadata descriptions published by the Services Catalog component can be derived from the capabilities of the distinct services that are depicted in the middle of the figure.

As it can be observed, there exists a close relationship between the different types of metadata used by the distinct GI services. In fact, starting from the metadata that accompany the geospatial data on the right side part of the figure, it could be possible to derive more or less automatically the rest of metadata descriptions.

1.6 Ontologies and Knowledge Representation in the context of Spatial Data Infrastructures

Ontology as a branch of philosophy deals with "the nature and the organization of reality" (Guarino and Giaretta, 1995). In philosophy, an ontology is a theory about the nature of existence, of what types of things exist; and ontology as a discipline studies such theories (Berners-Lee et al., 2001). But the term ontology has been used more recently in information systems and knowledge representation to denote a knowledge model, which represents a particular domain of interest. A body of formally represented knowledge is based on a conceptualization: the objects, concepts, and other entities that are assumed to exist in some area of interest and the relationships that hold among them (Genesereth and Nilsson, 1987). A conceptualization is an abstract, simplified view of the world that we wish to represent for some purpose. And the term ontology is used in this knowledge representation context to denote "a explicit formal specification of a shared conceptualization" (Gruber, 1992; Borst, 1997). This means that the ontology is explicitly defined and there is a formal notation, interpretable by machines and that the conceptualization is accepted by a group. They provide a shared and common understanding of a domain that can be communicated across people and application systems.

Ontologies have been developed in Artificial Intelligence to facilitate knowledge sharing and reuse. As it is mentioned in (Denny, 2002), the current heir to the artificial intelligence legacy may well be ontologies. But nowadays, ontologies are also a popular research topic in various communities such as Natural Language Processing, Cooperative Information Systems, Intelligent Information Integration, Knowledge management or the incipient conception of the Semantic Web. In the semantic Web vision, unambiguous sense in a dialog among remote applications or agents can be achieved through shared reference to the ontologies available on the network.

Ontologies may vary not only in their content (the knowledge that are representing), but also in their structure and implementation. In (Denny, 2002) ontologies may be classified by: the level of description, the conceptual scope, the type of instantiation, and the specification language.

Concerning the level of description, how one goes about describing something reflects a progression in ontologies from simple lexicons or controlled vocabularies, to categorically organized thesauri, to taxonomies where terms are given distinguishing properties, to full-blown ontologies where these properties can define new concepts and where concepts have named relationships with other concepts, like "changes the effect of" or "buys from". This last category of full-blown ontology coincides with (Berners-Lee et al., 2001), who state that the most typical kind of ontology consists of a taxonomy and a set of inference rules. The taxonomy defines classes of objects and relations among them. Classes, subclasses and relations among entities are a very powerful tool for expressing knowledge. For example, an *address* may be defined as a sub-class of *location*, and a *city code* may be defined as a property which only applies to *addresses*. And inference rules (also called axioms) are rules to add semantics and to infer knowledge. They represent implicit knowledge about concepts and relations. For instance, an ontology may express the rule "if a city code is associated with a state code, and an address uses that city code, then that address has the associated state code".

As far as conceptual scope is concerned, ontologies also differ in respect to the scope and purpose of their content. The most prominent difference can be found between domain ontologies (specific files of endeavor) and upper ontologies (basic concepts and relationships invoked when information about any domain is expressed in natural language).

Another distinguishing property is the type of instantiation. Ontologies have two parts: the terminological component (like the schema for a relational database) and the assertional component (instances and individuals that manifest that terminological definition). Whether the 1965 Ford Mustang GT is an individual Ford automobile, or the vehicle with license plate number AXL429 is an individual Ford (as an instance of the subclass 1965 Ford Mustang GT), may vary between two valid automotive ontologies.

And last, ontologies may use a wide range of specification languages, which even includes general logic programming languages like Prolog. In fact, talking about ontology specification languages is almost equivalent to describe the history of knowledge representation techniques and their associated languages. Following, we will present some of the most relevant languages that support ontology construction.

- *Frame-based languages.* It is said that ontologies evolve from semantic networks (Quillian, 1967) notions, one of the earliest knowledge representation tools. Semantic Networks represent knowledge under the form of a labeled directed graph. Specifically, each node is associated with a concept, and the arcs represent the various relations between concepts. However, early semantic networks suffered from the drawback that they did not have clear semantics. The ambiguity in semantics arises from the fact that in Semantic Networks arcs can represent different kind of relations between nodes, basically because they make a blurred distinction between intensional knowledge (relations between concepts) and extensional knowledge (relations between individuals). Semantic networks were extended by frame-based representations (Minsky, 1981; Fikes and Kehler, 1985). According to (Schaerf, 1994), a *frame* usually represents a concept (or a class) and it is defined by an identifier, and a number of data elements called *slots*, each of which corresponds to an attribute that members of the class can have. The values of the attributes are either elements of a concrete domain (e.g. integers, strings) or identifiers of other frames. Additionally, each slot contains a series of descriptive properties informing about the corresponding attribute that are called *facets*. Examples of these facets are: default values; restrictions on possible fillers; and attached procedures for computing values when needed or for propagating side effects when the slot is filled (commonly called daemons). A frame can also represent a single individual, and in this case it is related with the attribute *instance-of* to the frame representing the class of which the individual is an instance. In order to promote knowledge sharing and reuse, it was proposed an application programming interface called Open Knowledge Base Connectivity (OKBC) (Chaudhri et al., 1998) for accessing knowledge bases stored in knowledge-representation systems that incorporates all the features of the basic approaches to frame-based systems. It provides a uniform model based on a common conceptualization of classes, individuals, slots, facets and inheritance. The protocol (with existing implementations in Common Lisp, Java and C) transparently supports networked as well as direct access to knowledge bases. Protégé (Stanford University School of Medicine, 2004; Noy et al., 2000) and Ontolingua (Farquhar et al., 1996) are examples of two OKBC-compatible tools that are widely used for ontology construction.
- *Description Logics* (Baader et al., 2003)[9]. Frame-based systems have been usually criticized in the literature (Schaerf, 1994) due to the same problem of semantic networks: their semantics was not completely defined (in particular the distinction between the frames links). Thus, other formalisms were searched to provide systems with an explicit model-theoretic semantics. This is the case of a set of languages based on a form of logic thought to be especially computable and known as Description Logics. Descrip-

---

[9] http://dl.kr.org/

tion Logics unifies and gives a logical basis to the well known traditions of frame-based systems, semantic networks, KL-One-like languages (KL-One (Brachman and Schmolze, 1985) was the predecessor of Description Logics), object-oriented representations, semantic data models, and type systems. A Description Logic model is based on the notions of *concepts*, which represent classes of objects with similar characteristics, *individuals* which are instances of concepts and roles which are relationships between individuals. Central to a Description Logic model are the notions of *subsumption* and *classification*. One concept is said to be subsumed by another when all of its instances are necessarily instances of the subsumer. The computation of the subsumption relation (by means of first-order logic proof methods) allows the automatic construction of a classification hierarchy, with conceptual definitions being arranged from the general to the specific. This automatic classification is precisely the advantage with respect to frame-based languages, where classification must be explicitly given by the designer of the knowledge base. Nowadays, one of the most famous languages derived from this family is Classic (Borgida et al., 1989).

- *Knowledge Interchange Format (KIF) (Genesereth and Fikes, 1992) and its successor Common Logic (CL) (Common Logic Working Group, 2004).* KIF arose as a standard format for knowledge exchange among different knowledge-representation systems (Classic, KL-One, ...). It is a monotonic first-order predicate calculus with a simple syntax and support for reasoning about relations. An example of a tool using the syntax and semantics of KIF is the Ontolingua Server. This may seem contradictory because Ontolingua (OKBC-compatible) offers a frame-based interface (through the Web) for the edition, browsing, translation and re-use of ontologies. But internally, it translates all the information into KIF.

- *F(rame)-Logic (Kifer et al., 1995).* F-Logic accounts in a clean and declarative fashion for most of the structural aspects of object-oriented and frame-based languages. In addition to this, this deductive object-oriented language is suitable for defining, querying and manipulating database schemas. OntoEdit [10] is an ontology editor based on this language.

- *General-purpose languages with declarative features.* Informally, declarative programming involves stating *what* is to be computed but not necessarily *how* it is to be computed. In declarative programming languages, knowledge takes the form of data that is managed by a general interpreter. Prolog and Description Logics, frames-based languages or rule-based systems are examples of declarative programming. But there are also several general-purpose languages (more industrial languages like C, C++ or Java) that have added some declarative features. And the description of this additional knowledge can be considered as another form of specifying an ontology. For instance, CLIPS (C Language Integrated Production

---

[10] This tool has been developed by Ontoprise GmbH (http://www.ontoprise.de).

System) (Giarratano and Riley, 1998)[11] is a tool for building expert systems which is written in C language and integrates objects with rules. The object model of CLIPS is inspired in CLOS (Common Lisp Object System) (Keene, 1989), a language that incorporates object-oriented features to Common LISP. CLIPS has also an equivalent Java version called JESS (Java Expert System Shell) (Friedman-Hill, 2003)[12]. The Protégé ontology editor, despite being considered as a frame-based tool, uses the CLIPS text file format as a default save/load file format for the definition of both domain classes and instances.

- *Ontology languages for the semantic Web.* It is also worthwhile mentioning again that the great impact of the Web has encouraged the use of ontologies. Nowadays, many ontology languages rely on W3C technologies like RDF-Schema (RDFS) as a language layer, XML Schema for data typing, and RDF to assert data. There are systems like Sesame (Broekstra et al., 2002) that provide the necessary infrastructure for storing and expressive querying of large quantities of data in RDF and RDF Schema. Another proposal compatible with RDFS is the Ontology Inference Layer (OIL) (Decker et al., 2000). It provides a web-based representation and inference layer for ontologies, which combines the widely used modeling primitives from frame-based languages with the formal semantics and reasoning services provided by Description Logics. This language has been also used as the basis for defining DAML+OIL (van Harmelen et al., 2001), a semantic markup language for Web resources which is currently being evolved into the Ontology Web Language (OWL) (Bechhofer et al., 2004) standard. DAML+OIL is an initiative sponsored by the DAML (DARPA Agent Markup Language) program of DARPA (the U.S. Defense Advanced Research Projects Agency), which began in August 2000 with the goal of developing a language and tools to facilitate the concept of the Semantic Web. DAML+OIL provides a rich set of constructs with which to create ontologies and to markup information so that it is machine readable and understandable.

As far as geographic information and spatial data infrastructures are concerned, it must be said that this research community is also aware of the potential benefits of using ontologies as a knowledge representation mechanism, which facilitates knowledge sharing and reuse in interoperable environments. In particular, three main areas for the application of ontologies have been identified within this multidisciplinary context of spatial data infrastructures:

- Data sharing and Systems Development. Works like (Pundt and Bishr, 2002; Visser et al., 2002) consider ontologies as the adequate methodology to support geographic data sharing. Ontologies help to define the meaning of features contained in geo-spatial data and they can provide a "common

---

[11] http://www.ghg.net/clips/CLIPS.html
[12] http://herzberg.ca.sandia.gov/jess/

basis" for semantic mapping. Other works like (Fonseca, 2001; Fonseca et al., 2000) go even further and propose the creation of software components from diverse ontologies as a way to share knowledge and data. These software components are implemented as classes derived from ontologies, using an object-oriented mapping. The use of an ontology, translated into an active information system component, leads to ontology-driven information systems, in this case ontology-driven geographic information systems.

- The own structure of metadata schemas can be considered as ontologies, where metadata records are the instances of those ontologies. In turn, this use of ontologies may be applied to the following aspects:

  - Ontologies may be used to profile the metadata needs of a specific geospatial resource and its relationships with the metadata of other related geospatial resources. For instance, collections of geospatial resources such as the items conforming a mosaic of ortho-imagery share a great percentage of metadata. Ontologies could be used to model the metadata elements that describe the collection at an upper level and the metadata elements that are specific to the items.

  - Interoperability across metadata schemas. Transformations of metadata between two different standards could be resolved by systems that observe the commonalities of the two ontologies and automatically detect the metadata element mappings. An example of an ontology-based interoperability solution is presented in (Weißenberg and Gartmann, 2003), where an ontology architecture is used to offer personalized Geo-Services to athletes, journalists and spectators in Olympia 2008. Different metadata standards are used to describe the different geo-services. These metadata standards (e.g. ISO 19115 and Dublin Core) are modeled as ontologies using F(rame)-Logic and semantic technologies are used to match these ontologies and enable semantic queries.

- As a tool for the classification of resources and the improvement of information retrieval. Metadata enhance information retrieval because they intend to describe unambiguously information resources. But this improvement depends greatly on the quality of metadata content. One way to enforce the quality of metadata is the use of a selected terminology for some metadata fields in the form of thesaurus or lexical ontologies. As mentioned in (Bechhofer and Goble, 2001), thesauri are useful in bridging the gap between the metadata provided by the metadata creator and the concepts presented by the searcher. Furthermore, this work remarks the importance of knowledge representation techniques in the creation of coherent thesauri: better than a related set of terms, thesauri should have an underlying ontology enabling the reasoning about concepts. Despite not being a GI specific project, an example of the use of ontologies in this area is the Healthcybermap project (Boulos et al., 2001), which combines metadata and ontologies to provide new ways of finding health information resources. There, the resources are marked-up with metadata or indexed in a central database using metadata. And explicit concepts in the metadata are mapped onto an ontology (e.g. a clinical terminology or classification or a collection of merged ontologies) allowing a search engine (Semantic Web agent) to infer implicit meanings not directly mentioned in either the resource or its metadata.

1.7 Conclusions

Spatial data infrastructures provide the framework for the optimization of the creation, maintenance and distribution of geographic information at different organization levels (e.g., regional, national, or global level) and involving both public and private institutions. As a consequence of this, Governments start considering spatial data infrastructures as basic infrastructures for the development of a country. Spatial data infrastructures are becoming as relevant as classical infrastructures like utilities (water, electricity, gas), transport or telecommunication infrastructures. And the creation of these infrastructures should follow a set of common strategies that makes possible the coordination among different initiatives.

On the other hand, this chapter has remarked that the development of spatial data infrastructures must take into account the background provided by multiple disciplines. In particular, digital libraries is a discipline that can offer a conceptual base, especially in technological aspects, for building spatial data infrastructures. This experience has its extension to geographic information through geolibraries. This type of digital libraries, specialized in geographic information resources, provides a very important know-how which can be used as starting point for creating concepts, processes and methods.

Additionally, this chapter has studied the use of metadata within spatial data infrastructures. Apart from presenting an overview of the different types of metadata and their standardization initiatives, it has been shown how metadata is the essential piece to interconnect the technical components of a spatial data infrastructure. The metadata describing the data and services offered by a spatial data infrastructure are closely related. In fact, the metadata describing the geospatial data holdings can be used to derive the metadata describing the capabilities of the services (Web Map Server, Web Feature Server, etc.) that provide access to these holdings. And similarly, metadata describing services are used as entries to the services catalogs (registries) that publicize the range of services offered by the spatial data infrastructure.

Finally, this chapter has also introduced the concept of ontology within the context of geographic information and spatial data infrastructures. Ontologies are used as a means to facilitate knowledge sharing and reuse. And as well as in many other communities (e.g., Natural Language Processing, Cooperative Information Systems, Intelligent Information Integration, or the Semantic Web), they are a popular research topic in geographic information systems and spatial data infrastructures. Furthermore, they provide an important basis in the proposals that will be presented in following chapters of this book.

A metadata infrastructure for the management of nested collections

2.1 Introduction

As regards the cataloguing of geographic resources, an important circumstance to take into account is the existence of collections or aggregation of geographic resources (or datasets) that can be considered as a unique entity. Most of these collections arise as a result of the fragmentation of geographic resources into datasets of manageable size and similar scale. In this sense, for example, the *Spanish National Geographic Institute (IGN)*[1] offers distinct versions of its products (*Cartographic Numeric Base BCN*, *National Topographic Map MTN*, *Digital Terrain Model MDT*,...) according to different scales: *BCN200* identifies the *BCN* at 1:200,000 scale; *BCN25* identifies *BCN* at 1:25,000 scale and so on. Each product-version pair compiles the set of files into which the Spanish territory was divided so as to provide, at the scale required, a number of files with reasonable size. Those files are usually named "tiles" and the *IGN* establishes for each scale the numbering and spatial extent covered by these tiles. Another example in the Spanish sphere is the Military Cartography from the Army. There, the term "series" is used to denominate the cartography (altimetry, milestones in railway and road networks, etc.) which is offered at different scales. For instance, the "L series", which is considered as the essential series of Spanish Military Cartography, gather the set of tiles which represent exactly the same extension as the ones of *National Topographic Map* at 1:50,000 scale. Besides, each aforementioned product may be composed of several information layers. For example, each *BCN* tile is composed of the following thematic layers: administrative divisions; altimetry; hydrography and coasts; buildings and constructions; communication networks; utilities; and geodetic vertexes. Finally, it is also frequent the use of the term "mosaic of ortho-imagery" to designate the set of files that compose the ortho-image of a geographic area. Although the aforementioned examples belong to the Spanish geographic information context, similar examples can be found in other

---

[1] http://www.ign.es/

countries and organizations like the *French National Geographic Institute*[2], the *U.S. Geological Survey*[3], the *Canadian Geospatial Data Infrastructure*[4] or the *Australian Agency for Geospatial Information*[5].

As it can be observed, the terminology to denominate the different types of geographic information collections is quite diverse. However, regardless of terminology, it is possible to distinguish two main types of collections: single-type collections and multiple-type collections FGDC (2002).

- A single-type collection represents the aggregation of multiple data units which were originated in similar conditions (same capture equipment, scale, ...) and with equivalent semantic content. Besides, each unit represents a geographic information piece that a user can order without requiring special processing to generate it by the geographic information provider. For instance, a collection of all monthly average sea surface temperatures (left side of figure 2.1) or a mosaic of digital ortho-photography (right side of figure 2.1) are considered as single-type collections. According to the distribution pattern followed by the components/units of the collection, these collections are usually classified into: spatial collections (components follow a preestablished spatial division), temporal collections or temporal series (components of the collection are obtained following a preestablished periodicity), and spatio-temporal collections (components of the collection are both periodically and spatially distributed).

- On the opposite, the second type of collections (multiple-type collection) compiles data layers or components coming from different sources in order to perform a GIS study or project. For example, a study of the effects of *El Niño* and *La Niña* events on vegetation could contain two information layers: *TOPEX/Poseidon* [6] total monthly average sea surface heights; and values for the *Normalized Difference Vegetation Index* (a model to transform satellite measures into superficial vegetation types) taken from *NOAA/NASA Pathfinder AVHRR Land Program*[7].

Finally, it is also common to organize resources in more than one level of aggregation, originating nested collections. By nested collections it is meant that a collection can be included as a part of another collection. This recursive definition of collections enables the hierarchical organization of resources in a repository.

As it has been mentioned in previous chapters, when providers or distributors of geographic information want to publish the content of their holdings,

---

[2] http://www.ign.fr/

[3] http://www.usgs.gov

[4] http://www.geoconnections.org/

[5] http://www.ga.gov.au/

[6] Web site of the TOPEX/Poseidon satellite at the Colorado Center for Astrodynamics Research (CCAR): http://www-ccar.colorado.edu/research/topex/html/topex.html.

[7] http://daac.gsfc.nasa.gov/CAMPAIGN_DOCS/LAND_BIO/GLBDST_main.html

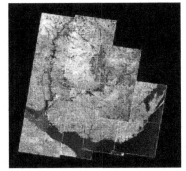

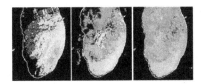

Temporal Data Series
Sea Surface Temperature (SST)
of Lake Michigan.

Mosaic of ortho-images
Landsat Multispectral Images of Uruguay

Fig. 2.1. Examples of single-type collections: temporal collections (left) and spatial collections (right)

they must provide standardized descriptions of their datasets (metadata), which are later incorporated into data catalogs and clearinghouses. The creation and maintenance of geographic metadata is a time consuming and thorough process. This circumstance is especially problematic if a collection of thousands of datasets must be documented. On one hand, the datasets belonging to the same collection share a high percentage of meta-information that must be replicated multiple times. And on the other hand, users of geographic information are accustomed to manage the entire collection as a unique entity (e.g. the *National Topographic Map* at scale 1:50,000), which should return by data catalogs as a unique result instead of displaying the complete list of thousands of files that conform the collection. Therefore, the cataloguing of each individual file separately seems to be not very recommendable. The meta-information must be replicated indiscriminately for each file and this process is likely to be error prone.

The problem of how to describe collections within metadata is an important issue in new proposals for geographic information metadata standards (e.g., ISO 19115 ISO (2003a) or Remote extensions of CSDGM FCDC (2002)). As it is stated in ISO (2003a), the notion of cataloguing a set of related documents together in a discoverable series is common in map catalogues. There, it is proposed that metadata can be derived for a series of related spatial datasets, and such metadata is generally relevant or can be inherited by each of the dataset instances. Thus, most of these metadata standards define elements to point at related resources, usually by means of a string or number conforming to a formal identification system.

However, a catalog system can not manage collections just enabling librarians to manually edit the fields concerned with these links. There are several aspects that justify a more complex implementation of collections.

- Firstly, the resources (and metadata records describing them) must be uniquely identified, at least within the local catalog. Thus, all the references among the aggregate and the parts must be always up-to-date whenever a component of the aggregation is added or removed.
- Secondly, the components that form part of a collection usually share a high percentage of meta-information (e.g., abstract, topic category, etc.). There are metadata elements whose content could be inherited from the metadata record that describes the collection. But if the catalog does not provide an automatic mechanism to inherit meta-information, metadata creators must replicate common descriptions for each dataset. For instance, using again the example of *TOPEX/POSEIDON* data, the only difference of meta-information between two datasets taken at different instant times is precisely the value corresponding to the metadata element "creation date".
- Thirdly, some values of the metadata elements in the collection metadata record are aggregated or averaged over the values of the components of the collection. Typical examples of these elements are the temporal extent or the spatial extent of the collection.
- And finally, as long as discovery services is concerned, it is accepted the relevance of presenting the user an aggregated view of what it is available instead of an infinite list of results (e.g., similar scenes/data available at several instant times) Longley et al. (2001).

The objective of this chapter will be to provide a metadata solution to manage nested collections in catalog systems, which is based on XML technologies and concepts derived from knowledge bases. The most accepted way to exchange metadata is by means of XML documents, whose syntax is enforced by control files in the form of DTDs or XML-Schemas. Thus, a system managing metadata records as XML documents and the syntax of those documents as XML-Schemas will be highly independent of the structure of metadata standards. This chapter proposes the construction of catalog services over a knowledge base component, which is able to store the different types of metadata schemas supported, the aggregation relations established among these schemas, and the inference mechanisms that these relations will provide.

In addition to this introduction, the remainder of this chapter is structured as follows. Next section revises approaches that have dealt with the problem of whole-part relationships in different contexts. Section 2.3 presents the desired functionality of a system able to manage collections. Then, section 2.4 explains the design of a Metadata Knowledge Base component, which is the base for the construction of a catalog able to manage collections. Section 2.5 will describe how a set of preestablished prototypical aggregation relations may facilitate the process of defining a collection scenario. And finally this chapter will end with some conclusions and future lines.

2.2 Related work

This section presents the work dealing with the concept of aggregations of resources in three main areas. Firstly, subsection 2.2.1 explains how different metadata standards (with special emphasis in geographic metadata standards) have defined elements that make references to related resources and collections. Secondly, subsection 2.2.2 introduces the problem of collections management from the perspective of digital libraries and geolibraries. And thirdly, subsection 2.2.3 revises the definition of structural relations in the knowledge representation field.

2.2.1 Addressing collections and relations in metadata standards

In general, the concept of referencing to related resources is an important issue in all types of metadata standards. For instance, Dublin Core DCMI (2004), a minimalist high-level metadata standard comprising, includes a *relation* element among its fifteen basic elements. This element is defined as "a reference to a related resource" and the recommended best practice is to identify the referenced resource by means of a string or number conforming to a formal identification system. Furthermore, despite the simplicity of the standard, the Dublin Core Metadata Initiative proposes eleven refinements for this element (see table 2.1). Ten of these refinements (two by two) correspond to the roles of five binary relationships which may be established between two resources. And although only the roles *isPartOf* and *hasPart* define explicitly the whole-part relationship of a typical collection, the rest of relationships determines a grouping of related objects. The definition of a collection, overall in the more general context of digital libraries, is open to many interpretations. For instance, multiple occurrences of the *hasVersion* refinement could be used to point at the collection of resources that are derived from the described resource.

In Remote Sensing extensions of the CSDGM FGDC (2002), metadata definitions have been added to describe the component parts of an aggregation or describing the larger aggregation of which a data unit or aggregation is a member, to allow the user to determine the level of aggregation to which a metadata element applies. Figure 2.2 shows the fragment of the production rules for the *Identification_Information* section of the extended standard, which support the concept of aggregation. A new *Dataset_Identifier* element allows unique identification of dataset, interpreted to refer to an aggregation of data at any level as appropriate to the context. Based on this unique identification an *Aggregation_Information* subsection enables the description of this dataset as being a component of a higher-level (*Container_Packet_ID*);

Table 2.1. Refinements of Dublin Core *relation* element

Name	Definition
isVersionOf	The described resource is a version, edition, or adaptation of the referenced resource. Changes in version imply substantive changes in content rather than differences in format.
hasVersion	The described resource has a version, edition, or adaptation, namely, the referenced resource.
isReplacedBy	The described resource is supplanted, displaced, or superseded by the referenced resource.
replaces	The described resource supplants, displaces, or supersedes the referenced resource.
isRequiredBy	The described resource is required by the referenced resource, either physically or logically.
requires	The described resource requires the referenced resource to support its function, delivery, or coherence of content.
isPartOf	The described resource is a physical or logical part of the referenced resource.
hasPart	The described resource includes the referenced resource either physically or logically.
isReferencedBy	The described resource is referenced, cited, or otherwise pointed to by the referenced resource.
references	The described resource references, cites, or otherwise points to the referenced resource.
isFormatOf	The described resource is the same intellectual content of the referenced resource, but presented in another format.
hasFormat	The described resource pre-existed the referenced resource, which is essentially the same intellectual content presented in another format.
conformsTo	A reference to an established standard to which the resource conforms.

or being composed of lower-levels (*Component_Information*) according to an *Aggregation_Criteria*.

This extension is oriented to catalog temporal series containing satellite imagery available at several different times. That is to say, it is targeted to single type collections that aggregate data components originated from a single source and which probably differ in one or a few metadata values.

The ISO 19115 standard document also remarks that there is a potential hierarchy of re-usable metadata that can be employed in implementing a metadata collection. By creating several levels of abstraction, a linked hierarchy can assist in filtering or targeting user request for metadata presentation to the requested level of detail. Following paragraphs will detail some parts of the model that are related with this linked hierarchy.

On one hand, the UML (Unified Model Language) (Booch et al., 1998) class diagram in figure 2.3 shows the hierarchy of geographic information classes to which metadata may apply in ISO 19115. Metadata is optional

```
Identification_Information = Dataset_Identifier + Citation +
  Description + Time_Period_of_Content + Status + Spatial_Domain +
  0{Processing_Level}1 + Keywords + 0{Platform_and_Instrument_Identification}n
  + [Band_Identification|Thematic_Layer_Identification] +Access_Constraints +
  Use_Constraints + (Point_of_Contact) + (1{Browse_Graphic}n) +
  (Data_Set_Credit) + (Security_Information) +(Native_Data_Set_Environment)
  + (1{Cross_Reference}n) + 0{Aggregation Information}n
  ...

Aggregation_Information = (1{Container_Packet_ID}n) + 0{Component_Information}1

Container_Packet_ID = Dataset_Identifier

Component_Information = 1{Aggregation_Member_ID}n+ 1{Aggregation_Criteria}n

Aggregation_Member_ID = Dataset_Identifier
```

Fig. 2.2. CSDGM Remote Sensing extensions to support aggregations

for the upper level of the hierarchy (*DS_Aggregate* class), which is defined as a collection or series of spatial data sharing similar characteristics (theme, source date, resolution, or methodology). These series are specified (subclassed) as: a typical dataset series (*DS_Series*), e.g. a collection of raster map data captured from a common series of paper maps; a general association (*DS_OtherAggregate*), e.g. a cross reference or larger work citation; or a special activity (*DS_Initiative*), e.g. a project, campaign or study. But in most cases, metadata usually applies to a dataset (*DS_Dataset*), which is defined in this context as a consistent spatial data product instance that can be generated or made available by a spatial data distributor. A dataset may be a member of a data series and may be also composed of a set of features and attributes, which in turn could have their own metadata associated. However, *DS_XXX* classes shown in figure 2.3 are external entities to the content of the metadata, that is to say, they do not appear inside a metadata file.

The first real approximation of ISO 19115 to describe the hierarchical relationship between two metadata files/records are the attributes *fileIdentifier* and *parentIdentifier* of the *MD_Metadata* class (root class of ISO 19115 metadata model, see figure 2.4). The attribute *fileIdentifier* is used to identify the current metadata file (describing a unit of a collection) and *parentIdentifier* can be used to reference the identifier of the collection metadata file. Here, it is worth mentioning the availability of a cataloguing tool that facilitates the implementation of this parent-child referencing. This tool, called M3Cat (see appendix C.1), enables the replication of metadata at the moment of creation of a child dataset because it copies automatically the values from the metadata elements of the parent dataset. But if the metadata of parent dataset is changed later, no synchronization is performed to update the values of child datasets. Additionally, it must be observed that by using *fileIdentifier* and *parentIdentifier*, datasets can only belong to one collection because there is only a single *parentIdentifier* in *MD_Metadata*. However, it could be also interesting to consider that a dataset may be bundled in different collections.

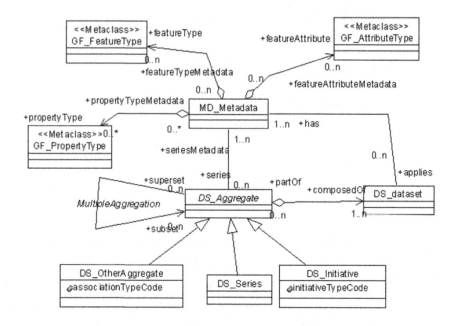

Fig. 2.3. Metadata application in ISO 19115

A second approach of ISO 19115 to manage the problem of aggregations is the definition of an *MD_AggregateInformation* class. This class informs about the whole-part relationships between datasets and aggregates inside metadata. The *MD_AggregateInformation* class provides information about the master dataset (aggregate) of which the dataset described by this metadata is a part. Figure 2.5 shows this class and its navigability from the *MD_Metadata* class. Similar to previous approaches, it stores identifiers to hierarchically superior elements. The *associationType* attribute describes the association type of the aggregate and the *initiativeType* attribute informs about the type of initiative under which the aggregate was produced. The definition of the code-lists used by these attributes is shown in table 2.2.

The main conclusion from above standards is that the meta-information about links between resources (datasets) is becoming very important. However, the implementation of these standards in a cataloguing system cannot be a simple (manual) edition of the fields concerned with this links. Some aspects that justify the complexity of implementation are:

- The resources (or metadata records describing them) must be uniquely identified, at least within the local catalog. And all the references among aggregate and parts must be always up-to-date. For instance, using CS-DGM Remote Extensions, whenever a dataset a is added to an aggregate b,

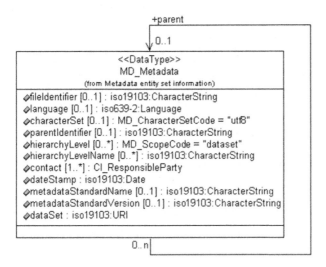

Fig. 2.4. Attributes *fileIdentifier* and *parentIdentifier* in *MD_Metadata*

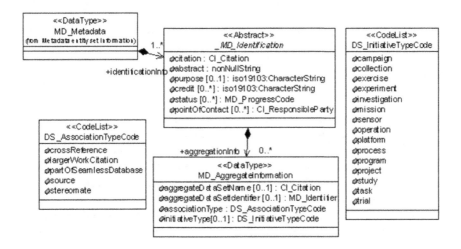

Fig. 2.5. *MD_AggregateInformation* class

container information of a must reference to b, and component information of b must be updated to include the reference to a.

- The datasets that form part of an aggregation usually share a high percentage of meta-information (e.g., abstract, topic category, etc.). There are metadata elements whose content could be inherited from the metadata

Table 2.2. Definition of codes for *associationType* and *iniativeType*

Value	Definition
DS_AssociationTypeCode	*justification for the correlation of two datasets*
crossReference	reference from one dataset to another
largerWorkCitation	reference to a master dataset of which this one is a part
partOfSeamlessDatabase	part of same structured set of data held in a computer
source	mapping and charting information from which the dataset content originates
stereoMate	part of a set of imagery that when used together, provides three-dimensional images
DS_InitiativeTypeCode	*type of aggregation activity in which datasets are related*
campaign	series of organized planned actions
collection	accumulation of datasets assembled for a specific purpose
exercise	specific performance of a function or group of functions
experiment	process designed to find if something is effective or valid
investigation	search or systematic inquiry
mission	specific operation of a data collection system
sensor	device or piece of equipment which detects or records
operation	action that is part of a series of actions
platform	vehicle or other support base that holds a sensor
process	method of doing something involving a number of steps
program	specific planned activity
project	organized undertaking, research, or development
study	examination or investigation
task	piece of work
trial	process of testing to discover or demonstrate something

record that describes the aggregate. One possible solution to avoid this replication could be to fill in these common elements only for the aggregate metadata record. Nevertheless, these elements may be mandatory in the standard and they should be in the metadata records describing the parts. And on the other hand, when a user wants the description of a single dataset, he would rather obtain a complete description with inherited and specific meta-information. Therefore, if the catalog does not provide an automatic mechanism to inherit meta-information, metadata creators must replicate common descriptions for each dataset.

- Some values of the metadata elements in the aggregate metadata record are aggregated or averaged over the values of the datasets that form part of the aggregation. Typical examples of these elements are the temporal extent or the spatial extent of the aggregate.

Other metadata standards have avoided the problem of referencing and inheriting among metadata records by using uniquely a metadata record for the whole collection. In this approach, the metadata standard provides some elements to specify the particular details of the components of the aggregation.

This is the case of the metadata that describes the capabilities of Web Map Servers (WMS) Beaujardière (2002). WMS is an OGC standard specification for servers producing maps of geo-referenced data on demand over the web. A typical use of this service is to provide views of satellite imagery that may be available at several different times and several different wavelength bands. For the purpose of description, *WMS* supports a *getCapabilities* operation to obtain service-level metadata (also named as capabilities) describing the content and acceptable request parameters of a *WMS*. Regarding the example, the general description of data received from satellite but belonging to different dates will differ only in time dimension. It makes no sense (or is not very efficient) to update daily the capability file of *WMS* in order to express that today it is able to serve a new layer of satellite imagery. It would be desirable to indicate inside the capabilities of a *WMS*, that it is able to serve any map representing one of the components of the aggregation.

Since version 1.1 of WMS specification, it is possible to characterize within server capabilities the multi-dimensional properties of the source data, including time, elevation and other sample dimensions. Thus multidimensional properties are published in service metadata, and a dimension parameter may be included in map request operations. Capabilities of a Web Map Server are usually implemented as an XML file conforming to a Document Type Definition (DTD). In the new specification the *Dimension* element (tag $< Dimension >$ in XML files) is used to declare that one or more dimensional parameters are relevant to the information holdings of that server. A *Dimension* element includes a required *name*, a required measurement *units* specifier, and an optional *unitSymbol*. The *Dimension* element is defined as follows in the specification DTD:

```
<!Element Dimension EMPTY>
<!ATTLIST Dimension name
  ID #REQUIRED
  units CDATA #REQUIRED
  unitSymbol CDATA #IMPLIED>
```

All dimensions in a *getCapabilities* XML response are server-defined, with two exceptions: the dimensions named *time* and *elevation*, which are privileged predefined special cases given their frequent use. Finally it must be remarked that the *Dimension* element does not provide valid values for a *dimension*; that is the role of the *extent* element. Having declared the existence of a dimension, the *getCapabilities* XML response uses corresponding *extent* elements to specify the bounds of a geodata object along zero or more independent single dimensions. The format of *Extent* element in XML is the following:

```
<Extent name="dimension_name" default="default_value"
  multipleValues="0|1" nearestValue="0|1">
  extent_value
</Extent>
```

The *extent_value* string declares what value(s) along the *Dimension* axis are appropriate for this specific geospatial data object (its syntax is also shown in table 2.3). Finally, figure 2.6 shows an example of the definition of a < *Layer* > with *Extents* in WMS Capabilities XML.

Table 2.3. Syntax of *extent_value*

Syntax	Meaning
value	A single value
value1, value2, value3, . . .	A list of multiple values
min/max/resolution	An interval defined by its lower and upper bounds and its resolution. A resolution value of zero means infinitely-fine resolution
min1/max1/res1, min2/max2/res2, . . .	A list of multiple intervals

```
<Layer>
<!-- Declare dimensions in use. Declarations are inherited by enclosed Layers. -->
  <Dimension name="time" units="ISO8601" />
  <Dimension name="temperature" units="Kelvin" unitSymbol="K" />
  <Dimension name="elevation" units="EPSG:5030" />
  <Layer>
    ...
    <!-- Specify extent of Layer. Extents are inherited by enclosed Layers. -->
    <Extent name="time" default="2000-10-17">
      1996-01-01/2000-10-17/P1D
    </Extent>
    <Extent name="elevation" default="0">0/10000/100</Extent>
    <Extent name="temperature" default="300">230,300,400</Extent>
  </Layer>
  ...
</Layer>
```

Fig. 2.6. Definition of a *Layer* within the capabilities of a *WMS*

The approach taken by WMS specification has proven to be a simple and efficient solution for managing aggregations whose components only differ in just a few dimensions. However, geographic data catalogs (as components of geolibraries and spatial data infrastructures) may require the management of more complex types of aggregations, e.g. aggregations whose components neither share the same coordinate reference system nor have been originated from the same source. Furthermore, we might be interested in managing recursive levels of aggregations.

2.2.2 Collections in Digital libraries and Geolibraries

The problem of managing collections of related resources is not new in the context of traditional libraries and digital libraries, informally defined as the electronic version of the first ones.

As mentioned in (Hill et al., 1999), traditional library collections are firmly associated with library holdings: a collection is a set of copies of materials that a library holds, just as a museum collection consists of the objects held by the museum. When libraries access information outside of their own holdings, they are conceptually accessing other collections and services. Exceptionally, a library collection may contain a number of special collections, such as rare books or maps, which are given special treatment. Another feature of these collections is that they usually imply the availability of a specialized service. According to (Lagoze and Fielding, 1998), collection development serves three important roles:

- The selection of resources that are members of the collection. These may be the whole contents of the library or a subset of the total resources.
- The specialization of discovery aids or cataloging techniques, which are tailored to the characteristics of the collection. Examples of these discovery aids are inventories, registers, indexes or guides that are created to provide detailed information of the repository, generally in hardcopy format.
- And the administration of the collection. This includes a set of management and preservation policies.

However, as it is stated in (Hill et al., 1999), the concept of a digital library collection is as broad as a dictionary definition: "a group of objects"; and the objects in a digital library are not necessarily physically owned by the library. A collection may be seen as a set of metadata records pointing to local resources, distributed resources or even (as with gazetteers or directories) to the real world. And the process of determining which object grouping must be treated as a collection is open to many possibilities: a set of objects sharing a uniform characteristic (e.g. topic, format, source, temporal coverage or geographic coverage) established by digital library managers; objects from various collections selected for their relevance to a current project; sets of query results saved for future reference; or metadata items selected by an information retrieval filter or agent.

A precedent of the management of collections in digital libraries can be found in the world of online bibliographic databases. Usually, these databases publish user guides that can be considered as collection metadata. An example of this are the Dialog Bluesheets[8]. Dialog is a company that developed in 1966 (prior to the era of Internet) one of the first online information retrieval system to be used globally with materially significant databases. Bluesheets are written guides for every database accessible through the Dialog service. They contain detailed instructions on search techniques for the special features of

---

[8] http://www.dialog.com/

each database, including file description, subject coverage, date range, up-date frequency, sources of the data, and the origin of the information. On the Bluesheet you will also find a sample record that shows what you can expect to obtain when you perform a search in the database.

On the other hand, as traditional libraries gave public access to their cat-alogs via the Internet, several standardization initiatives appeared to describe the contents of a collection:

- The Encoded Archival Description (EAD) (U.S. Library of Congress, 1998) standard was created for encoding archival finding aids to collections of materials. As exchange format, it uses Standard Generalized Markup Lan-guage (SGML), the markup language from which XML derives.
- A special profile of Z39.50 information retrieval protocol was created for the access to digital collections (U.S. Library of Congress, 1996). This pro-file provides search and retrieval services by means of descriptive records (metadata records), which are classified into two categories: collection de-scriptive records (provide an overall description of a collection as well as collective or individual descriptions of some or all of the objects in the collection), and object descriptive records (describing digital objects of physical objects).
- The Research Support Libraries Programme (RSLP) Collection Descrip-tion Project developed a model allowing all the projects in its program to describe collections in a consistent, machine readable way (Powell et al., 2000). One remarkable feature of this model, based on the Dublin Core schema, is that it has attributes for expressing relationships between col-lections: *Sub-collection*, *Super-collection* and *Associated collection*. This model has been used by the SMETE Digital Library[9] for the definition of virtual library collections (Geisler et al., 2002).

A relevant work to facilitate the access to digital library collections is the STARTS protocol (Gravano et al., 1997). This protocol for internet re-trieval and search, developed by the Stanford University, facilitates the task of querying multiple document sources, namely text collections accessed via search engines. The existing search engines are typically incompatible be-cause of three main problems: they support different models and interfaces (the query-language problem), they do not return enough information with the query results for adequate merging of results (the rank-merging prob-lem), and they do not export metadata about the collections that they index (the source-metadata problem). The goal of STARTS is that the search en-gines implementing the protocol will assist a meta-searcher in choosing the best sources to evaluate a query, evaluating the query at these resources, and merging the query results from these sources. The basis for the implementa-tion of STARTS protocol is the availability of source metadata, describing the contents of the collection. This collection metadata consists of two pieces:

---

[9] http://www.smete.org/

- Source metadata attributes. It consists of a list a list of metadata attribute-value pairs, describing properties of the source. This includes information that a meta-searcher can use to rewrite the queries sent to the source as well as other attributes manually generated (e.g., abstract, contact or access constraints).
- Source Content Summary. This piece of metadata contains information that is automatically generated such as: list of words that appear in the source; statistics for each word listed; or total number of documents in source.

Lagoze and Fielding take a step further and present in (Lagoze and Fielding, 1998) a design for a digital library collection service which enables the introduction of structure into a distributed information space. The main contribution of this approach is that now the resources contained in the collection may be distributed across multiple repositories (each one having its own interface). Therefore, the collection is logically defined as a set of criteria for selecting resources from the broader information space. For instance, a collection may be defined as a query that restricts the value of subject metadata element to "computer science". This type of collections enables a dynamic growth of the collection from resources that appear in multiple repositories. Another feature of this service is that it is independent from other services and mechanisms in the digital library. This way, the collection service neither constrains other organizational models nor does it impose structure when it is neither needed nor desired.

Within the context of geolibraries (digital libraries filled with geographic information), a good example of a system dealing with the problem of collections is the Alexandria Digital Library (ADL) project (Janée and Frew, 2002; Goodchild and Zhou, 2003). In ADL, collection metadata is used to model collections and give support for both computer processing and human use (Hill et al., 1999). ADL collections of geographically referenced items (maps, aerial photographs, satellite images, recordings, etc.) are described by means of:

- Collection level metadata. This is a standardized description about the collection. Collection Level Metadata focuses on the aggregated information about collections and description of collection services such as search and discovery. And similar to source metadata of STARTS protocol, it makes a clear distinction between inherent metadata (automatically generated) and contextual metadata:
  - Inherent metadata information can be derived through the computer analysis of the contents of any collection such as: item counts by format or type of item; histograms describing spatial and temporal coverage; or the types of geospatial footprints (e.g., points, bounding boxes); or an example of the full metadata content of an item.
  - Contextual metadata is information that must be supplied by the collection provider or collection maintainer, it cannot be otherwise derived

from the collection contents. These contextual metadata includes elements like: the title; the responsible party; scope and purpose; the type of collection (digital items, off-line items, gazetteer, etc.); the query parameters (if it is a result set from a query); or the special behaviors (e.g., search semantics) that the collection may exhibit or require in specific operational contexts (e.g., when accessed by a particular search engine).

- Item level metadata. This level of metadata compiles the individual descriptions of the items that form part of the collection. Item-level information includes an identifier that is unique within the collection.

The main contribution of the ADL system with respect to previous approaches is its geographic-oriented approach. Unlike text-oriented approaches (STARTS, bibliographic databases, etc.) it has identified the relevance of presenting geographic characteristics of the collections such as the visualizations of geographic and temporal coverage. This is interesting not only for maps, remote-sensing images or aerial photographs, but also for the rest of media resources, which might be contained in digital library collections. The drawback of this solution is, however, that it does not enable the nesting of collections. The collections defined above have only one level of aggregation. The cause of this may be the strict separation between item metadata and collection metadata, using in most cases different metadata schemas. It seems logical that collection metadata needs more metadata elements to describe the characteristics of the collection but in essence a collection should be treated in the same way as other resources contained in the library.

Another relevant work in the context of geographic information is the implementation of the NASA's Global Change Master Directory (GCMD) described in (Vogel and Northcutt, 1999). The goal of GCMD is to enable users to locate and obtain access to Earth science data sets and services relevant to the global change and Earth science research. Here, the resources are organized into nested collections (in this case called directories), which in last term aggregate a list of atomic units. The GCMD allows incremental information retrieval through the use of parent (i.e. generalized) and child (i.e. specific) metadata records. The association between parent and child records is accomplished through a metadata attribute in the children. Metadata records are created according to DIF (Directory Interchange Format) and the attribute *Parent_DIF* in child metadata record contains the identifier of the parent record. As a first step only the information about upper level directories is displayed to the user. And if this information appears to be of interest, the user may decide to retrieve the specific information of one of the atomic units. In order to perform this first step discovery, the SQL search of the GCMD database identifies parents when other records (i.e., children) exist that point to it. Only information in the parent is searched, ignoring the child records. Thus, each parent must include information from the children, in order for the user to be able to ultimately find a child through its parents. This is pre-

cisely the cause of one of the main disadvantages of this system: the creation
of parent metadata records require a large amount of human investment. As
mentioned in (Vogel and Northcutt, 1999), this could be avoided by means
of automated reverse inheritance, i.e. the parents inherit information of its
children.

2.2.3 Addressing relations in knowledge representation

In knowledge representation, structural relations are used to structure knowledge in groups of concepts. Since the introduction of Quillian semantic networks (Quillian, 1967), taxonomical links have been commonly used in the
representation of knowledge. But apart from the taxonomic relation (the *is-
a*), other relations have been defined to individuate, refine, or structurally
aggregate concepts.

An especially relevant work in the definition of relations is the classification introduced in (Sathi et al., 1985). There, Sathi et al. present a theory of
activity representation which is based upon a layered representation of knowledge. It consists of the five following layers (from the lowest to the highest
one): the implementation layer with primitives for machine interpretation of
the concepts and the assertions; the logical layer that defines the word concept
as a collection of assertions; the epistemological layer that provides a way of
regulating the flow of information through inheritance and other structural
relations; the conceptual layer which is comprised of models of common primitives (e.g. concepts of time, activity, state, agent, ownership, etc.), reused
across domains; and the domain layer to provide concepts, words, and expressions specific to a domain of application. The interesting point in this work is
that the epistemological layer structures knowledge in six relations to provide
defaults, classification, aggregation, elaboration, revision and *individuation*.
Besides, it specifies what information may be transferred between two related
concepts. This layered representation uses a frame-based language and the
relations are modeled by means of distinguished slots in the schemas (frames)
as follows:

- *Defaults.* This relation is the relation *is-a* but reduced to the role of the
  definition of default properties, i.e. assignment of the default properties
  through the *is-a* relation. A schema (or frame) can inherit all the slots
  (except for the *is-a*) and values along the *is-a* relation.
- *Classification.* This relation represents the process by which a set is divided
  or partitioned into subsets on the basis of some attribute values. The slot
  has-subset is used to relate a set to its subsets; and *subset-of* is the inverse
  of *has-subset*. In terms of semantics, all slots (except for *subset-of*) and
  values can be inherited across the *subset-of* relation.
- *Elaboration.* It represents the process by which a concept is expanded and
  filled in with details. The *has-elaboration* slot relates a prototype to the
  detailed individuals; and *elaboration-of* is the inverse of *has-elaboration*.

All slots (except for *elaboration-of*) and values can be inherited along the *elaboration-of* slots.

- *Aggregation*. This relation represents the combination of parts to make a whole. The slot *part-of* is used by the disaggregates to point at the aggregate concept. The inverse of *part-of* is *has-part*. As concerns inheritance, parts inherit some attributes from their aggregation (e.g. ownership), and on the other hand the aggregate concept may aggregate (e.g., cost) or average (e.g., performance) other attributes.

- *Revision*. This is the process of deriving a new object from an original object by adding some improvements. It represents a transformation process, a revision in time. As with other relations, revision may be viewed from the two sides: the *revision-of* slot is used by the derived object to point at the original object; *revised-by* is the inverse link. A schema containing the *revision-of* slot can introduce revisions by adding or transforming slots.

- *Individuation*. It represents the development of the individual from the universal. It can be interpreted as a copy of a prototype with an individual name and exceptions, if any. The instantiated schema uses the *instance* slot to point at the prototype.

The contribution of this categorization of relations has been the identification of information that can be inherited across the relation. It must be also remarked that some relations (*classification, elaboration, aggregation* and *revision*) need two slots to identify the different roles played by the concepts at each side of the binary relation. That is to say, the inheritance of information is different depending on the role (played). For instance, a schema inherits all slots and values across *subset-of*, but in contrast *has-subset* enables no inheritance. In fact, the own slots can be considered as ten individual relations, being these relations asymmetric (if $a \, \Re \, b$ is true, $b \, \Re \, a$ is not true), transitive, and having eight of them their inverse relation.

(Artale et al., 1996) also remark that knowledge bases, data bases and object-oriented systems (referred as Object-Centered systems) all rely on attributes as the main construct used to associate properties to objects; and among these, a fundamental role is played by the so-called *part-whole* relation. They state that the representation of such structural information requires a particular semantics together with specialized inference and update mechanisms, but rarely do current modeling formalisms and methodologies give it a specific, "first-class" dignity. This paper presents some formalisms adopted in knowledge representation (e.g. extensions of Description Logics) and object-oriented systems. But perhaps the most remarkable feature of this work is the revision of the research done about *part-whole* relations in linguistic and cognitive studies. Particularly relevant it is the distinction among various kinds of specialized *part-whole* relations presented in (Winston et al., 1987):

- *Component/Integral-Object*: Integral objects are characterized by having a structure, while their components are separable and have a specific functionality. For example, "Wheels are parts of cars".

- *Member/Collection*: It captures the notion of membership in a collection. For instance, "A tree is part of a forest".
- *Portion/Mass*: The whole is considered as a homogeneous aggregate and its portion are similar to it and separable, as in "This slice is part of a pie".
- *Stuff/Object*: It expresses constituency of things and can be paraphrased using is *partly* or *is made of*, as in, "The bike is partly steel".
- *Feature/Activity*: It designates a phase of an activity. A phase, like a component, has a functional role but it is not separable.
- *Place/Area*: It is a spatial relation among regions occupied by different objects. For example, we can say that "An oasis is part of a desert".

In the sense of improving the expressivity and semantics of object-oriented models, it is also worth-while mentioning the work of (Zarazaga-Soria, 2000). This thesis provides a framework for the reutilization of C++ code, which is based on the meta-information of object oriented models. The C++ language is extended with features derived from frame-based languages in order to include knowledge in the object-oriented language. In this framework attributes are extended with *facet*-like meta-information, which facilitates typical tasks in information systems such as persistence (special facets describe the name, type and constraints of table columns where attribute values must be stored) or user interface (facets describe how attribute values must be edited and presented to the user). Additionally, as a special kind of attribute, relation attributes are defined to access transparently the objects participating in a relation. In fact, the relation itself is modeled as a special class having knowledge about the structure of the participating classes; and these special classes are specialized by the relation cardinality.

2.3 Defining the desired functionality of a collection enabled catalog system

Figure 2.7 shows the hierarchical structure of metadata describing a resource produced by the Spanish National Geographic Institute (IGN): the Cartographic Numeric Base at 1:200,000 scale (BCN200). The BCN200 contains core geographic data (administrative divisions, altimetry, hydrography and coasts, buildings and constructions, communication networks, utilities, and geodetic vertexes) in digital format, which is later used to derive hard-copy maps at this scale. This resource can be considered as a collection that groups the files providing real data for each province in Spain. Due to the lack of space, not all the metadata elements for the description of the collection have been displayed in figure 2.7. But in contrast, this figure displays all the elements that may differ for the description of the components in the collection. And as it can be observed they are not very numerous: just the specific *title* of the component; the *reference date*; the *geographic location identifier* (code and name of province); the *bounding box* that defines the spatial extension covered by each component; the *coordinates reference system*; and the URL of the *online resource*.

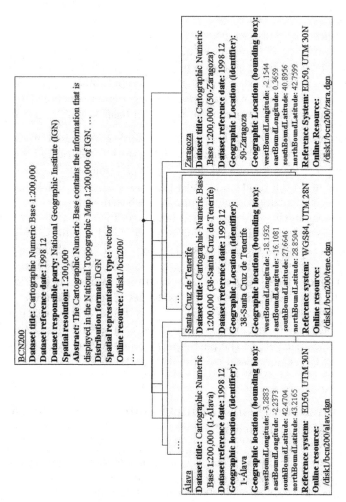

Fig. 2.7. Cartographic Numeric base

The case of BCN200 is not an isolate example. In fact, according to the type of collection (e.g., spatial collections aggregating components that are spatially distributed to cover a wide area, or temporal series aggregating resources with similar characteristics but taken at different instant times), it is possible to imagine the metadata elements that will probably differentiate the description of two components in the same collection. That is to say, instead of creating complete descriptions of each component in the collection manually, a system could automate this labor just having a high-level description of the entire collection and the specific values of just a few elements for each component. Table 2.4 shows the elements belonging to the "ISO 19115 Core metadata for geographic datasets" profile and the subset of those elements that are frequently redefined according to some prototypical types of collections. The redefined elements are marked with an x under column redefined (R). For instance, for a spatial collection usually divided into tiles, the following elements could be redefined at component level: the *title* of the unit which frequently includes the numbering of the tile; the *dataset reference date* element which stores the date of creation or publication of the tile; the *responsible party* element because sometimes the construction of tiles are subcontracted to different companies; the geographic *bounding box* element; the *reference system* element because some tiles may use a different coordinate system or different parameters (e.g. UTM zone); or the *online-resource* element which defines the way to access the geospatial data. Besides, this table also remarks some elements that are subject to be summarized and stored as common elements at collection level (marked with x under column S). For example, for spatial and spatio-temporal collections, it results interesting to calculate the minimum *bounding box* that covers the *bounding boxes* of the components.

Table 2.4. Elements redefined (R) at component level and summarized (S) for collection level

Core metadata element	Spatial		Temporal		Spatio-temporal	
	R	S	R	S	R	S
01. Dataset title (M)	x		x		x	
02. Dataset reference date (M)	x		x		x	
03. Dataset responsible party (O)	x		x		x	
04. Geographic location of the dataset (by four coordinates or by geographic identifier) (C)	x	x			x	x
05. Dataset language (M)						
06. Dataset character set (C)						
07. Dataset topic category (M)						
08. Spatial resolution of the dataset (O)						
09. Abstract describing the dataset (M)						
10. Distribution format (O)						
11. Additional extent information for the dataset (vertical and temporal) (O)			x	x	x	x
12. Spatial representation type (O)						
13. Reference system (O)	x					
14. Lineage statement (O)						
15. On-line resource (O)	x		x		x	
16. Metadata file identifier (O)						
17. Metadata standard name (O)						
18. Metadata standard version (O)						
19. Metadata language (C)						
20. Metadata character set (C)						
21. Metadata point of contact (M)						
22. Metadata date stamp (M)						
Count	6	1	5	1	6	2

The conclusion is that we would like to have a system able to synthesize metadata descriptions for the entire collection and the components, thus avoiding the redundancy of metadata. Additionally, this system should give support for the management of nested collections (recursive levels of aggregations between resources). In contrast to the ADL proposal (Janée and Frew, 2002) where a special schema for collection level metadata was created, our system should manage all resources, collections or atomic units, in the same way. On the other hand, this system should also provide enhanced presentation services of collections including the generation of histograms describing spatial and temporal coverages or item statistics by type, formats, and so on. Finally, it should be remarked that this catalog system should be not constrained to a specific metadata standard or schema. The system should be enough general to manage collections described by Dublin Core, ISO 19115 or other metadata standards.

Figure 2.8 shows the expected metadata pieces to be supported in the desired system. The left side of the figure shows how a typical metadata system would catalog the datasets compounding the collection (described by MD_is) and the collection itself (described by
$MD\_Collection$). And on the right side of the figure, it is shown how the different metadata records on the left side can be synthesized (compacted) into a minimized set of metadata pieces. These minimized pieces of metadata include: $MDS\_collection$, metadata which is common at a collection level; MDS_is, pieces of metadata containing the specific or redefined metadata element values for each unit in the collection; and the *characterization* of the collection. The idea is that there exists a biyective function that relates the individual metadata descriptions (on the left side) to the compacted description (on the right side). Look at appendix A.1 for a discussion about the consistency properties that must exhibit a compacted description of collections and components.

The $MD\_Collection$ represents metadata that could have been created by a librarian just analyzing the characteristics of the collection. Following the approaches of ADL Collection Level Metadata (Janée and Frew, 2002) or STARTS protocol (Gravano et al., 1997), we could divide $MD\_Collection$ into:

- Contextual metadata. Elements that provide a general description of the collection, and that must be supplied by the maintainer of the collection. These metadata includes generic descriptions such as *abstract, topic category* or *spatial representation type* of the components in the collection.
- Inherent metadata. This part includes elements whose values are summarized from the specific descriptions of the components. An example of this second type could be the geographic *bounding box* element whose value is the envelope or minimum bounding box that covers the specific bounding boxes of the units. These elements are obtained as a result of applying

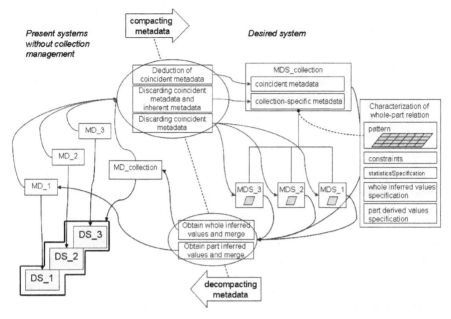

Fig. 2.8. Metadata pieces in the desired system

specific aggregated functions for those elements that are redefined at unit level.

As one of the desired objectives of the system is the minimization of metadata,
MDS_Collection in figure 2.8 should store uniquely contextual metadata of the collection. In addition, it is possible to make two subtle distinctions among the contextual metadata elements stored in *MDS_Collection*:

- *Coincident metadata.* These are the elements are specially intended to avoid replication in the components. That is to say, if the complete description of components (*MD_1, MD_2, MD_3*s) were available, the values of these elements would have been obtained by observing coincidences of metadata element values. For instance, if the distribution format of all components were TIFF (Tagged Image File Format), the distribution format would be uniquely stored at collection level metadata.
- *Collection-specific metadata.* This category includes the elements whose value has been filled in specifically to describe the context of the collection. As mentioned in (Marshall, 1998), a collection is likely to be more than an accretion of all it contains; it has been gathered for a purpose. Collection-specific metadata is thus vital for articulating the scope, intent, and function of a particular collection (attributes that are likely to make the collection easier to locate, and easier to use). For instance, collection-

specific metadata includes a *purpose* element with the goal under which the collection was created, or an *abstract* element describing the collection as a whole.

Whenever the complete description of the collection ($MD\_Collection$) is needed, the system would reconstruct the inherent metadata and would add it to $MDS\_Collection$. This inherent metadata would be computed by applying a series of aggregated functions over the metadata elements of the components. This aggregated functions are specified in the *whole inferred values specification* of the *characterization*.

MDS_i represents the minimum meta-information that should be stored for each component assuring no information loss with respect to MD_i. In other words, if the complete description of components (MD_is) were available, each MDS_i would have been obtained as the result of discarding the coincident elements detected for $MDS\_Collection$. Additionally, other elements values could be skipped or minimized in case there were a function that could derive the value from other elements of MDS_is and $MDS\_collection$. For instance, the title of the components in the IGN products can be derived from the concatenation of the generic title of the product and the code of the specific component (e.g., the numbering of a tile). In this case, the storage of the specific code of the component would suffice. This would not represent an information loss because this system should be able to reconstruct the MD_is when needed. For that purpose, a merging process should be applied to MDS_i and $MDS\_Collection$. To obtain the value of an element e, this process should take into account the following cases ordered by priority:

- Firstly, check whether there exists a function to derive the value of e. For instance, there may be a function for the *title* element that obtains the value by concatenating the title of the collection and a special code. Another example would be the case of a repeatable element like *keywords*. A function may be specified to add the keywords contained in $MDS\_Collection$ and the keywords in MDS_i. These functions are specified in the *part inferred values specification* of the *characterization*.
- Secondly, check whether there is a value for e in MDS_i.
- And thirdly, check whether a value for e can be found in $MDS\_Collection$. This case could be considered as a case of inheritance by default.

Finally, the *characterization* depicted in figure 2.8 should store the special features of the collection (some of them have been already mentioned):

- *Pattern.* Identification of a spatial/temporal pattern that may follow the components in the collection. Components of spatial collections usually follow some type of prefixed division of the space. Knowing this pattern will facilitate documentation and organization of the components in the collection. Additionally, it enables the supervision of the status of cataloguing of the collection. For instance, working with maps divided into tiles, the status of cataloguing could be supervised by means of a visualization tool that overlaps two layers: a coverage that establishes the extension of tiles; and a layer consisting of the bounding boxes of the components, which have been already catalogued.

- *Constraints.* There are possible constraints that the metadata records describing the components in the collection must observe. For instance, let us take the case a collection that aggregates a series of components spatially distributed over a concrete area. In such scenario, it is recommendable to impose that all metadata records describing the components should include a valid *geographic location* element (in the form of a bounding box, a geographic identifier, or other types of location references).

- *Statistics specification.* Specification of statistics that may be interesting to have a general idea of the collection, e.g. item counts by format or type of item or histograms describing spatial and temporal coverage.

- *Whole inferred values specification.* This is the specification of the functions that automatically generate inherent metadata elements for the description of the collection.

- *Part derived values specification.* This is the specification of the functions that enable the merging of some elements of MDS_is and $MDS\_Collection$ to derive some values for MD_i.

2.4 The Metadata Knowledge Base

2.4.1 Building the catalog services over a metadata knowledge base

Figure 2.9 shows the architecture of the catalog system that will be able to support the management of nested collections of resources. Apparently, the interface offered by the *Catalog Server* component depicted in figure 2.9 does not differ very much from that defined in section 1.5. The *Catalog Server* component offers a set of services that support the management, discovery and access of resources by means of a series of metadata entries that describe these resources. On one hand, the management services are usually accessed by client applications like metadata editors to organize the catalog entries in a local repository. On the other hand, the discovery services allow users (e.g., catalog clients using standardized interfaces or customized search interfaces incrusted in Web Portals) to search among these metadata entries using an established query language. And finally, the access to the resources is redirected through the pointers included in the metadata entries describing them. Additionally, the catalog controls the access to their services by means of user accounts and associated sessions.

However, what marks the difference between typical catalog systems without collection support and our collection-enabled catalog system can be found in the way of handling the repository of metadata. Frequently, catalog implementations use relational databases for the storage of metadata entries,

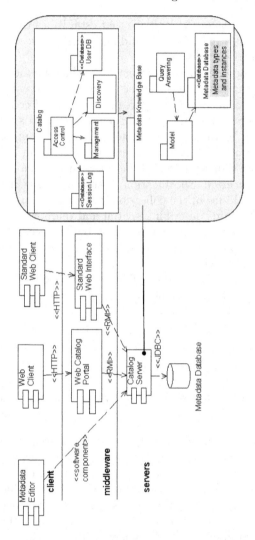

Fig. 2.9. Architecture of the collection-enabled catalog system

making profit of SQL (Structured Query Language) for the implementation of discovery services. Nevertheless, an ad-hoc and direct implementation of catalog services over a relational database does not seem the best solution for our objectives. According to the expected functionality described in section 2.3 the catalog system must offer the following features: it should enable flexible definition of metadata records (probably not constrained to a specific metadata standard); it should support recursive levels of aggregations (i.e., nested collections), enabling the registration of relations between collection metadata records and component metadata records; and it should provide inference mechanisms between relations established between metadata records.

In order to deal with all these prerequisites, this chapter proposes the development of catalog services over a *Metadata Knowledge Base* component. A Knowledge Base System is defined as a system that includes a knowledge base about a domain and programs that include rules (inference mechanisms) for processing the knowledge and for solving problems relating to the domain. As mentioned in (Borgida and Brachman, 1993), Knowledge Management Systems are being used in a variety of situations where access is needed to large amounts of data stored in existing relational databases. And this is a similar scenario: our catalog may manage thousands or even millions of metadata records that, in the long term, must be stored in a relational database. But instead of accessing directly the database, the proposed Knowledge Base component provides substantive semantics and reasoning facilities to accomplish this work. For instance, apart from the concrete occurrences of metadata records, the knowledge base will store the definition of metadata standards managed by the catalog. In the last decade Geographic metadata standards have continuously evolved and each specific community may define its own extension or profile. But thanks to the knowledge base, the catalog will be scalable enough to support gradually new standards or their modifications without having to reconstruct the software.

Next subsections will be devoted to explain in detail this Knowledge Base component, which marks the difference to other existent catalog system implementations. In particular, we will make a special emphasis on the following features of this knowledge base:

- Section 2.4.2 will present the model that has been used to support the storage of knowledge representations, i.e. the definition of collection scenarios and concrete metadata instances. This model is able to access the Relational Data Base Management System which in last term enables the storage of metadata in a robust and consistent repository. However, the description of this component is beyond the scope of this work.
- Section 2.4.3 will present the capabilities for the automatic generation of metadata of this Knowledge Base component.
- And section 2.4.4 describes the query answering capabilities of the knowledge base, which result vital to facilitate the discovery services of the catalog.

2.4.2 The knowledge base model

The way to represent knowledge in this knowledge base could be based on the concept of ontology. As mentioned in section 1.6, an ontology is usually defined as an "explicit formal specification of a shared conceptualization" (Gruber, 1992). In the context of information systems and knowledge representation, the term ontology is used to denote a knowledge model, which represents a particular domain of interest. And more specifically in the context of metadata standards, the own structure of metadata standards (also called metadata schemas) can be considered as ontologies, where metadata records are the instances of those ontologies. Therefore, ontologies may be used to profile the metadata needs of a specific resource and its relationship with the metadata of other related resources. For instance, metadata standards like Dublin Core (general purpose metadata) and ISO 19115 (geographic metadata) have been modeled as ontologies using the Protégé ontology editor (Stanford University School of Medicine, 2004; Noy et al., 2000). Another example can be found in (Weißenberg and Gartmann, 2003), which describes a system that offers personalized Geo-Services to athletes, journalists and spectators in Olympia 2008. For that purpose, different metadata standards are used to describe the different geo-services (each one using a different metadata standard). There, metadata standards are modelled as ontologies using F(rame)-Logic (Kifer et al., 1995) and semantic technologies are used to match these ontologies and enable semantic queries.

Figure 2.10 shows the ontology representing the metadata needs for the collection and components of the BCN200 example presented in section 2.3 (see figure 2.7). In this case, a frame-slot-facet representation (Minsky, 1981) has been used to specify such ontology in a graphical way. There, each frame represents a different type of metadata schema. Although a metadata schema is usually structured in sections and subsections, it is assumed (in order to facilitate visibility) that these schemas can be simplified into a flattened list of elements abstracting us from their complexity. The slots displayed inside the frames correspond to the elements of the "ISO 19115 Core metadata for geographic datasets" (ISO, 2003a) (already displayed in table 2.4), for the sake of clarity not all the elements have been displayed. Besides, it can be observed that there are three types of relations between frames. The *is-a* hierarchy is used to create more specific metadata schemas which add more slots or modify the slots of the parent frame. The *whole-part* hierarchy is used to establish the relation between the metadata describing a collection and the metadata that describe the components belonging to that collection. And the *instance* hierarchy is used to relate instances of a metadata schema to the frame that establishes its syntax.

Another question that may arise from the model in figure 2.10 is why two different schemas, *MD_IGN_Collection* and *MD_IGN_Component*, should be created for the description of IGN collections and components. In principle, all metadata instances should follow the syntax imposed by *MD_ISO19115*, which represents the ISO 19115 standard. The answer to this question can be found in the different inference behavior of *MD_IGN_Collection* and *MD_IGN_Component* with respect to the *whole-part* relation. Depending on the position of a frame with respect to the *whole-part* relation, the frame will obtain the values of slots in different ways:

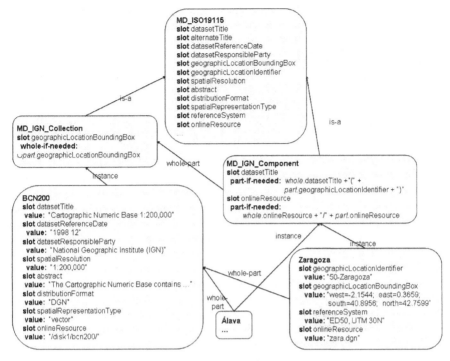

Fig. 2.10. Frame Model of BCN200

- The frames acting as *parts* (e.g. *MD_IGN_Component*) will obtain the value of a slot using one of the following prioritized ways:
    1. If the slot has a *part-if-needed* facet, this facet will be applied in first order. The *part-if-needed* is a daemon that returns a value obtained as the combination of slot values of the *part* and slot values of the *whole*. For instance, the *part-if-needed* of *datasetTitle* in figure 2.10 concatenates the *datasetTitle* of the *whole* and the *geographicLocationIdentifier* of the *part*.
    2. The second possibility is that the slot has its own value.
    3. And thirdly, if a slot has no value, the value of this slot is inherited by default through the *whole-part* relation.
- The frames acting as *wholes* (e.g. *MD_IGN_Collection*) will obtain the value of a slot using one of the following prioritized ways:
    1. If the slot has a value, this will be the final value.
    2. If the slot has not a value but there is a *whole-if-needed* facet, this facet daemon will be applied. This daemon is usually implemented as an aggregated function applied over the components of the aggregation. For instance, the *whole-if-needed* daemon of *geographicLocationBounding-*

Box in figure 2.10 computes the minimum bounding box covering the geographicLocationBoundingBox of the parts.

- Finally, if a frame is a whole and a part at the same time, it will first act as a part and then as a whole. Figure 2.11 shows the implementation of the if-needed facet daemon, which returns the final value of a slot ascending (if-needed-being-part) or descending (if-needed-being-whole) through the whole-part hierarchy. In order to avoid cycles, if a frame a is part of b, the following restrictions are applied: the values that a requires from b will be either direct values or values obtained by b acting as a part; and the values that b requires from a will be either direct values or values obtained by a acting as a whole.

```
exampleFrame

slot exampleSlot
  if-needed:
    if (if-needed-being-part provides a value)
    then return if-needed-being-part
    else if (if-needed-being-whole provides a value)
    then return if-needed-being-whole
      else return null
  if-needed-being-part:
    if (part-if-needed provides a value)
    then return part-if-needed
    else if (value is not null)
        then return value
        else if ( exampleFrame act as a part)
            then return whole.exampleSlot.if-needed-being-part
            else return null
  part-if-needed:
    combination of direct values (using value facet) of other slots
    in exampleFrame and values of whole slots returned by if-needed-being-part
  if-needed-being-whole:
    if (value is not null)
    then return value
    else if (whole-if-needed provides a value)
        then return whole-if-needed
        else return null
  whole-if-needed:
    aggregated function over the values obtained from slots of parts invoking
    if-needed-being-whole
```

Fig. 2.11. Inheritance behavior for a frame acting as part or whole

Although this frame-based solution seems to solve the problem of metadata duplication, the direct implementation by means of a frame-based language (understood in general terms as a knowledge-based approach) introduces important disadvantages:

- Historically, frame-based languages have not been enough exploited in industrial applications. An ontology management tool like Protegé, which is used in more than 100 countries and claims to be one of the most efficient tools, has not experienced with a real system containing more than 150,000 frames (classes & instances). However, a catalog managing collections (a sole spatial collection may contain more than 5,000 thousand of files) should manage the order of millions of metadata records. As mentioned in (Forbus and de Kleer, 1993), knowledge engineering specific tools present two main disadvantages: they are usually not appropriated in many contexts (e.g. not very efficient, not available in all platforms); and the state of art in reasoner engines is evolving continually.

- Secondly, using this frame-based solution, we need to define new frames not only for each metadata standard but also for each special behavior. Much of the functionality to infer meta-information through *whole-part* relation depends on the metadata standard. For instance, the metadata element that contains the geographic location (bounding box) of a resource is called *spatial* in Dublin Core and *MD_Metadata/identificationInfo/extent/ EX_Extent/geographicElement* in ISO 19115. Thus, in each standard a different *whole-if-needed* daemon would be needed to infer the minimum bounding box of a collection. On the other hand, the most accepted way to exchange metadata is by means of XML documents, whose syntax of this XML is enforced by control files in the form of DTDs or XML-Schemas. Given that standardization organizations usually publish these XML-Schemas and DTDs (e.g., ISO 19139 (ISO, 2003b) provides the XML-Schema for the implementation of ISO 19115 in XML), the question is clear: "Why must we rewrite this syntax in the form of frames or other concept-based representations?".

- One of the expected functionalities of the system was to provide collection statistics, which include histograms of spatial coverage or temporal coverage. The main application of this envisioned system will be in the geographic information world. Therefore, the system must facilitate the work with visualization tools and manage spatial data. However, frame-based languages do not provide many facilities for the work with complex data types.

Given these disadvantages, instead of using an existing knowledge-base software (e.g. Protégé, Classic (Borgida et al., 1989), etc.), we have opted for our own implementation of the knowledge base management system. This knowledge base management system has been developed following an Object-Oriented methodology (using Java as programming language) and its main features are that it reinforces the role of relations and that it makes profit of XML technologies. On one hand, works like (Artale et al., 1996) (already mentioned in section 2.2.3) encourage the improvement of semantics and inference mechanisms of *whole-part* relations in object-centered systems. In this case, our knowledge base enables the definition of *whole-part* relations where we have transferred the inference mechanisms previously found in the frames (*if-needed* facets). This way, frames are only focused in representing metadata, not in the behavior involved in *whole-part* relations. And on the other hand, the use of XML technologies increments the flexibility of the knowledge base. As mentioned before, most metadata is exchanged in XML files, whose

syntax is specified by XML-Schemas. Thus, a knowledge base managing metadata records (instances) as XML documents and the syntax (frames) of those
documents as XML-Schemas will be highly scalable and independent of the
particular structure of each metadata standard.

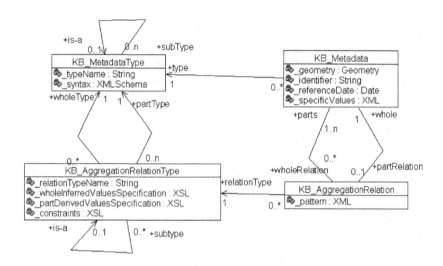

Fig. 2.12. Knowledge base

Figure 2.12 shows the main classes for the implementation of this knowledge base (for the sake of clarity, not all the methods and attributes attributes
are displayed). As it can be observed, there are two differentiate parts in the
model: on the left side, the classes that represent the metadata types and the
relation types (the knowledge); and on the right side, the classes that represent the instances of these types, i.e. the specific metadata records and their
relations.

The *KB_MetadataType* class represents the syntax of a metadata schema or
standard. It has a *syntax* attribute which stores the XML-Schema that defines
the syntax of a particular type of metadata. This class has a reflexive relation
(*is-a*), which is used to indicate that a metadata schema is an extension of
another schema.

The *KB_AggregationRelationType* class represents the types of relations
established between two metadata types. This class has three main features:

- The inference knowledge provided by the aggregation relations is specified
  in the attributes *_wholeInferredValuesSpecification* and *_partDerivedValues
  Specification*, which correspond to the *whole-if-needed* and *part-if-needed*
  daemons of the frame model (figure 2.10) respectively. The domain type

of these attributes is an XSL (eXtensible Stylesheet Language) (W3C, 2004a) document. XSL is a language for expressing style sheets that integrates a transformation language (XSL Transformations or XSLT) (Clark, 1999) which enables the definition of rules to transform an XML-document into another XML-document. See the discussion in section 2.4.3 about the selection of this domain type.

- The _constraints attribute stores the specification of the constraints (if applicable) that the components of the collection must observe. The domain type of this attribute is also XSL.

- This class has also a reflexive relation (is-a) to reflect inheritance between aggregation relation types. This reflexive relation facilitates the construction (definition of instances of KB_AggregationRelationType) of aggregation relation types. Thus, new aggregation relation types may inherit the constraints and inference specifications.

The KB_Metadata class represents instances of metadata which conform to a particular
KB_MetadataType. The specific (manually created) meta-information of a metadata record is stored in the _specificValues, whose domain type is an XMLDocument that should conform to the XML-Schema stored in the syntax attribute of KB_MetadataType. Exceptionally, this class includes two attributes, _geographicLocation and _datasetReferenceDate, that store the value of two metadata elements, which are also stored in _specificValues. The reason to have these redundant elements is to facilitate spatial and temporal queries and to speed up the generation of coverages.

And the KB_AggregationRelation class is used to describe the instances of the aggregation relations that are established between metadata records, provides attributes and methods which are common this group of metadata records that form part of a collection. This class includes a _pattern attribute to identify (if it is applicable) the default spatial/temporal pattern that follow the components. An example where these patterns appear would be the case of geographic information collections that have arisen as a result of the fragmentation of geographic resources into datasets of manageable size and similar scale. Usually, the spatial area covered by the components of these collections follow some type of prefixed division (e.g. the grid establishing the division of tiles for a specific scale, the province boundaries, etc.) of the space. Knowing this pattern will facilitate documentation and organization of the components in the collection. This pattern is particular of a relation instance. That is to say, one may define a KB_AggregationRelationType with an prototypical inference behavior that is reused in many collection scenarios where the only difference is the pattern. Since this pattern depends on the type of collection, the nature of this attribute may also differ enormously (e.g., a spatial coverage, a temporal frequency specification, or a list of keywords). Therefore the data type of this attribute is an XML-Document, which enables a flexible encoding.

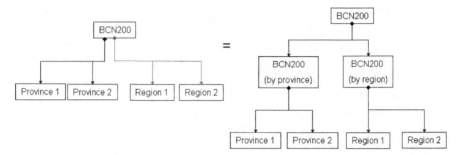

Fig. 2.13. Multiple relations associated to the same collection

Concerning the cardinality of the aggregation relation that this model establishes between metadata records, the following features must be remarked:

- An instance of *KB_Metadata* acting as *whole* can be only related, at maximum, to an instance of *KB_AggregationRelation* with the role *partRelation*. This means that all the metadata records describing the components of a collection share the same properties, those stored in the instance of *KB_AggregationRelation*. The necessity of having multiple occurrences associated with the same instance of *KB_Metadata* could be justified in the case of a collection organized by two or more different forms. For instance, figure 2.13 (left) shows how the BCN200 (Cartographic Numeric Base at 1:200,000) may be organized by provinces subdivisions or by regions subdivisions. But in essence, each type of organization provides the same data. Both the group of region files and the group of province files cover the Spanish territory. Here it arises the question whether we should allow these multiple relations. Although it seems interesting to distinguish these different types of grouping in a collection, it implies as well some disadvantages in order to generate inherent metadata. For instance, item statistics may result confusing as we do not know exactly how many the members of the collection are. On the other hand, we could avoid the problem of having distinct types of groupings by creating two separate subcollections, e.g. the first one compiling the province files and the second one compiling the region files (see right part of figure 2.13). Therefore, we finally restricted the association between *KB_Metadata* and *KB_AggregationRelation* to a 1:1 association.
- A metadata record may belong to more than one metadata collection. For instance, figure 2.14 shows again two possibilities of grouping the *BCN200* components. *BCN200 (by province)* collection aggregates the province files. *BCN200 (by region)* also aggregates the province files but there is an intermediate level of aggregation before accessing to the province files. In a first level, *BCN200 (by region)* aggregates a set of subcollections, which correspond to each Spanish region. And in a second level of aggregation,

each region subcollection aggregates the province files. And as it can be observed, the metadata records describing the leaf files are shared by two collections. For example *Province 1* is shared by *BCN200 (by province)* and by *Region 1*.

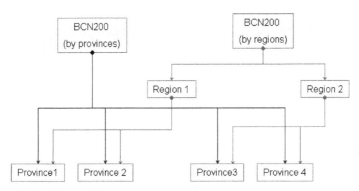

Fig. 2.14. A metadata record may belong to different metadata collections

As regards the storage of metadata records, this knowledge base makes use in the last term of a relational database. Details about the database model are beyond of the scope of this work. Nevertheless, the entity-relationship model of the database can be almost directly translated from the object-oriented model presented in figure 2.12. And with respect to the selection of the Data Base Management System (DBMS), we selected Oracle because of the availability of the Oracle Intermedia Text package, which facilitated the management of XML data in the database. Anyway, other DBMSs with XML support could have been selected.

2.4.3 Automatic generation of metadata

Metadata inference

With respect to the dynamic behavior of this model, the most important feature is the ability to infer complete metadata descriptions, ascending or descending through the aggregation relations. Figure 2.15 displays the methods of *KB_Metadata* and *KB_AggregationRelation* that provide this behavior, which is similar to the behavior already presented in figure 2.11 for frame facet daemons.

Let us see now some details about the methods involved in the automatic inference of metadata records:

- The method *getCompleteValues* act as the facet *if-needed* but providing the complete values for all the elements. It makes use of the methods

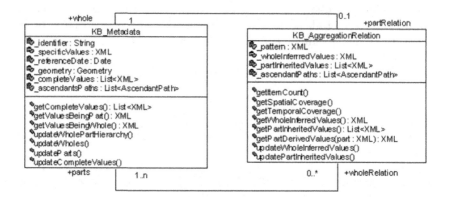

Fig. 2.15. Methods providing metadata inference

getValuesBeingPart and *getValuesBeingWhole*, which correspond to the *if-needed-being-part* and *if-needed-being-whole* facets respectively.

- The method *getValuesBeingPart* uses, in turn, the methods *getPartInheritedValues* and *getPartDerivedValues* to infer meta-information for a metadata record acting as part. Firstly, the *getPartInheritedValues* method enables parts to inherit meta-information contained in metadata records through the ascending *whole-part* hierarchy. And secondly, the *getPartDerivedValues* method enables a part to merge its metadata element values with the values obtained from *getPartInheritedValues* and according to the functions specified in the *_partDerivedValuesSpecification* of *KB_AggregationRelationType*.
- Finally, the method *getValuesBeingWhole* makes use of the method *getWholeInferredValues*, which obtains inherent metadata (metadata derived through the analysis of the components of the aggregation) according to the aggregated functions specified in the *_wholeInferredValuesSpecification* of *KB_AggregationRelationType*.

Appendix A.2.1 includes the algorithms (Java code) of the previous methods. Nevertheless, for the sake of clarity, all details with respect to database access (retrieval of XML stored in the database) have been obviated. Additionally, it must be mentioned that, for the sake of efficiency, the knowledge base stores some values that are automatically generated. This is the case of the attribute *_completeValues* in *KB_Metadata* and the attributes *_wholeInferredValues* and *_partInheritedValues* in *KB_AggregationRelation*. A collection may be composed of thousand of records and the inference process may slow down the system if all these values are recalculated whenever the *getCompleteValues* method is invoked. Besides, update modifications of metadata holdings are not so frequent as consultation.

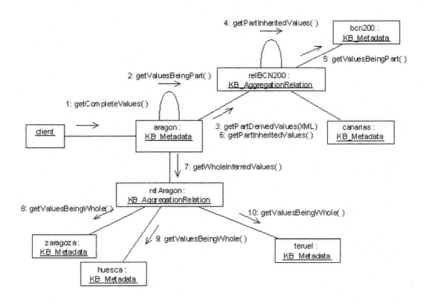

Fig. 2.16. Collaboration diagram for metadata inference

In order to illustrate this metadata inference process, figure 2.17 displays a detailed example of the metadata inferred for the BCN200 collection. This time the BCN200 has been organized in two levels of aggregation: at a first level BCN200 is composed of subcollections corresponding to the regions of Spain; and at a second level, each region subcollection is composed of the files corresponding to the provinces forming part of this region. It can be observed that metadata stored in the attributes _partInheritedValues and _wholeInferredValues of the relations are reused by different records to obtain their complete metadata. The figure remarks with italics the metadata that has been inferred for *bcn200*, *aragon*, and *zaragoza* record within the _completeMetadata attribute. Additionally, figure 2.16 displays the sequence of methods invocations to obtain the complete metadata of *zaragoza* object.

Additionally, it is worth mentioning the use of XSL as the domain type of _partDerivedValues Specification and _wholeInferredValuesSpecification in the KB_AggregationRelationType class. This was not our initial approach when we started defining the knowledge base. At that moment, we had decided to define our own language, based in XML, for the specification of these functions. Figure 2.18 shows a proposal definition of such a language where an *element-Name* tag is used to indicate the output element that the function (tag *rule*) should be returned according to a series of predefined functions (*add, concatenate, group*). However, the different nature of standards and the necessity of expressing functions that should operate over structured elements (e.g., union of bounding boxes or even polygons) incremented the complexity of this lan-

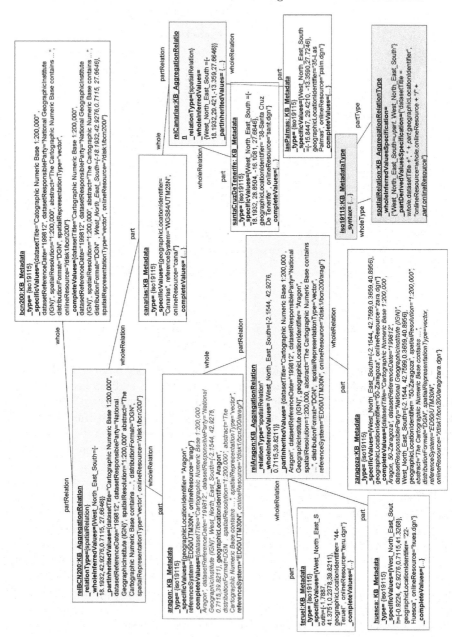

Fig. 2.17. Metadata inference

```
<!ELEMENT rules (rule+)>
<!ELEMENT rule (add | group | concatenate)>
<!-- add rule: as output all occurrences of the input element 'elementName' in the
    components are returned -->
<!ELEMENT add (elementName)>
<!ELEMENT elementName (#PCDATA)>
<!-- group rule: rule for whole-inferred-values.
    The value of the output element 'generatedElement' is obtained by the
    aggregation of the values of input elements 'elementName'.
    'elementName' may be a structured element that should be grouped by
    'groupByElement'.
  -->
<!ELEMENT group (elementName, groupbyElement?, generatedElement)>
<!ELEMENT groupbyElement (#PCDATA)>
<!ELEMENT generatedElement (#PCDATA)>
<!ATTLIST generatedElement operation (sum | max | min) #IMPLIED>
<!-- concatenate rule: the value of the output element is the concatenation
    of values specified by concatenatedElement (parent or child) -->
<!ELEMENT concatenate (elementName, concatenatedElement+)>
<!ATTLIST concatenate separator CDATA #IMPLIED>
<!ELEMENT concatenatedElement (#PCDATA)>
<!ATTLIST concatenatedElement hierarchyLevel (upper | lower) "lower">
```

Fig. 2.18. An initial DTD for the specification of functions

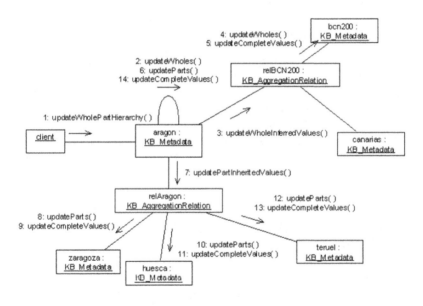

Fig. 2.19. Collaboration diagram for updating whole-part hierarchy

guage to support all the possibilities. Additionally, we needed to construct
an interpreter to perform the functions expressed in these XMLs. Finally, we
desisted from this approach because we realized that XSL already provided
the flexibility and syntax to process (without additional coding) XML inputs
and generate any type of output. Thus, it results ideal to specify the inference

that will combine or obtain values from the XML metadata of *whole* and *part* metadata records. The drawback is that the construction of XSL documents is not obvious for a catalog user but this is also true for the initial XML approach. In both cases, appropriate GUI interfaces must be provided for the specification of these functions. To exemplify the use of this type of XSL documents, appendix A.2.3 shows an XSL document that obtains a whole inferred value (the minimum bounding box) from a set of metadata records in conformance with ISO 19115 XML format. This stylesheets receives an XML input file, whose root element (called < *components* >) groups the XML of the set of individual metadata records (i.e., the metadata from the components of the collection, or the metadata from a parent and a child record) that must be processed. Further details about the functions that may appear in *_wholeInferredValuesSpecification* and *_partDerivedValuesSpecification* can be found in appendix A.1.

Finally, it must be mentioned that although the attributes *_completeValues*, *_wholeInferredValues* and *_partInheritedValues* improve the efficiency of metadata inference, they must be recalculated whenever the specific XML of a metadata record is modified or a record is inserted into a collection. The method *updateWholePartHierarchy* in *KB_Metadata* launches a series of updates in the whole-part hierarchy:

- Firstly, it invokes the method *updateWholes* to tell higher aggregation relations, that the precalculated *_wholeInferredValues* must be invalidated and recalculated again. It makes use of the *updateWholeInferredValues* method of *KB_AggregationRelation*.
- Secondly, it invokes the method *updateParts* to tell lower aggregation relations, that the precalculated *_partInheritedValues* must be invalidated and recalculated again. It makes use of the *updatePartInheritedValues* method of *KB_AggregationRelation*.
- And finally, once all the inferred values in the hierarchy have been recalculated, it invokes the method *updateCompleteValues* to to invalidate the *_completeValues* attribute and recalculate it again.

The collaboration diagram displayed in figure 2.19 shows the sequence of methods invocations to update the whole-part hierarchy of the *BCN200* (already presented in figures 2.17 and 2.16) starting from the *zaragoza* object. For further details, appendix A.2.2 contains the algorithms for the previous methods.

Generation of statistics

As a special case of automatic metadata inference, the *KB_AggregationRelation* class also offers methods to generate special statistics of the elements in a collection (see figure 2.15). These methods are the following:

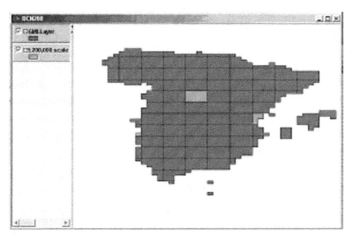

Fig. 2.20. Supervising the cataloguing status of the components of the collection

- The *getItemCount* method aims at generating statistics by type or format. At present it only returns the number of components of the collection. But in the future, more sophisticated statistics could be generated. Furthermore, as well as the metadata inference is specified by the attributes *_wholeInferredValuesSpecification* and *_partDerivedValuesSpecification* in *KB_AggregationRelationType*, another XSL attribute could be used to process the input XML of the component metadata records and generate an XML output containing the desired statistic.
- The *getTemporalCoverage* generates the temporal coverage of the components in the collection. As a special case, this method makes use of the attribute *_datasetReferenceDate* in *KB_Metadata*, which, despite being another metadata element, was specifically created to speed up the generation of this coverage.
- And the method *getSpatialCoverage* generates a spatial coverage of the components. This method also makes use of a special attribute called *_geographicLocation*, which was created to speed up the generation of the spatial coverage. Additionally, it must be remarked that this method returns a spatial coverage encoded in GML (Geographic Markup Language) (Cox et al., 2003). GML is an XML encoding for the transport and storage of geographic information, including both the spatial and non-spatial properties of geographic features[10].

Finally, it must be remarked that these statistics provide overviews of collections from different perspectives and that they may be used to supervise

---

[10] (ISO, 2003c) defines a geographic feature as "an abstraction of a real world phenomenon; it is a geographic feature if it is associated with a location relative to the Earth".

the status of cataloguing. For instance, in the case of spatial collections we could compare the spatial coverage with the pattern specified in the collection. Figure 2.20 displays a visualization tool that overlaps two layers: a pattern coverage in pink color that establishes the extension of tiles for a spatial collection at 1:200,000 scale; and a layer in green color that consists of the bounding boxes of the components already catalogued for the BCN200 example. In fact, the window presented in figure 2.20 has been taken from a metadata editor making use of this collection-enabled catalog. See chapter 5 for more details about the features of this metadata editor.

2.4.4 Intelligent query answering

A knowledge representation system should offer a number of reasoning services that can deduce implicit knowledge from that given explicitly by the designer of the knowledge base. For instance, the basic reasoning services that are typically carried out on concept-based knowledge bases are enumerated below. For a better understanding of these services, it must be mentioned that Concept-based knowledge bases use concept languages like Description Logics (Baader et al., 2003) which are based on first order logic semantics. As well as frame-based languages, concept languages are made of two components: a general schema concerning the classes of individuals to be represented (the classes of individuals are called concepts); and an instantiation of this schema, relating individuals to concepts.

- *Concept satisfiability.* Given a knowledge base and a concept (class/frame), does exist at least one model (occurrence) of the knowledge base assigning a non-empty extension to the concept? This is important not only to rule out meaningless concepts in the knowledge base design phase, but also in processing the user's queries, to eliminate parts of a query which cannot contribute to the answer. In languages like Description Logics queries are expressed as concepts defined by the user.
- *Subsumption.* Given a knowledge base and two concepts, is one concept more general than the other one in any model of the knowledge base? Subsumption detects implicit dependencies among the concepts in the knowledge base.
- *Consistency.* Are the schema of classes and the instantiation consistent with each other? That is, does the knowledge base admit a model?
- *Classification.* Where exactly is the concept situated in a concept hierarchy? Using subsumption, we can build a classification lattice of concept definitions. The classification is minimal with respect to subsumption relationships, thus if A subsumes B and B subsumes C there will be no direct link between A and C.
- *Instance checking.* Given a knowledge base, an individual and a concept, is the individual an instance of the concept in any model (occurrence) of the knowledge base.

- *Retrieval (or query answering)*. Given a concept, find all the objects occurring in the knowledge base that are instances of the concept.
- *Realization*. Given an individual occurring in the knowledge base, find the most specific concepts of which the individual is an instance.

Since the knowledge base presented aims at serving the needs of catalog services, we will only focus on reasoning services for retrieval. This is the objective of the *Query answering* component displayed in the figure 2.9, which shows the architecture of the catalog system.

But apart of retrieving the metadata records that verify the restriction specified by the user, the *Query Answering* component should provide with an intelligent query answering. Intelligent query answering is defined in (Cuppens and Demolombe, 1988) as the problem of "analyzing the intent of query and providing generalized, neighborhood or associated information relevant to the query". In particular, we are interested in providing an incremental retrieval of metadata records. That is to say, instead of overwhelming the user with an infinite list of results, an aggregated list of results would be more convenient for presentation purposes. This aggregated list of results would be obviously guided by the aggregation relationships between the records describing the components of a collection and the record describing the whole collection. Later, in a second step the user may explore the components of a specific collection verifying the restriction.

The intelligent detection of this aggregated list of results is a task that must be integrated within the *Querying Answering* process (see figure 2.21). This section focuses on this task of automatically collapsing results, but in order to illustrate the context of this task we will give some details about the steps of this workflow:

1. *Query Transformation*. The queries received from the user must conform to a query language with a recognized syntax. Several query languages have been proposed in the literature for catalog services such as: the OGC Common Query Language (defined within the OGC Catalog Interface Implementation Specification (Nebert, 2002)), which is a language similar to the specification of WHERE clauses in SQL; or RPN based languages (queries are expressed in Reverse Polish Notation format), which are used by the Z39.50 search and retrieval protocol (ANSI, 1995). However, we finally selected the Filter Encoding Specification (Vretanos, 2001), which is based on XML language and is widely used in last versions of OGC specifications. Thanks to the advances in XML technologies, this language results much more suitable for catalog implementation. Coming back to the problem of this first step, these queries must be transformed into a query tree, where the nodes represent logical operators (*and, or*) and the leaves contain SQL queries which will be later executed over the database, storing metadata records. In our prototype implementation of the knowledge base, the *WHERE* clauses of these SQL queries use the Oracle *CONTAINS* operator (similar operators can be found in other DBMS: e.g.,

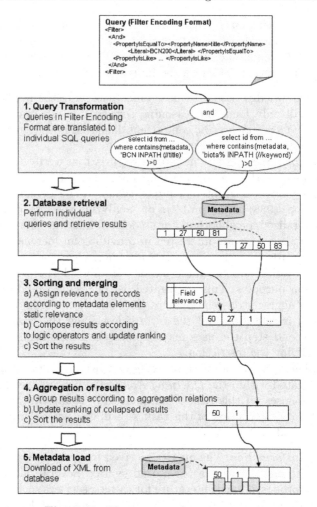

Fig. 2.21. The process of query answering

the *MATCH* and *AGAINST* functions of MySQL). This operator, together with an appropriate index, enables queries on columns containing Document Object Like Data (including XML documents). Besides, the *CONTAINS* operator allows the inclusion of XPath (Clark and DeRose, 1999) expressions for addressing parts of an XML document, i.e. filtering the restrictions over specific metadata elements. XPath is a language that provides the syntax for expressing regular path queries, the basic querying mechanism for semistructured data like XML (Calvanese et al., 2003).

2. *Database retrieval*. Secondly, the individual queries contained in the leaves of the aforementioned query tree must be executed to return a set of lists containing the records that verify the individual restrictions. These queries

make profit of text indexes over the columns that contain the metadata of each record in XML format.

3. *Sorting and merging.*
   a) Firstly, the records retrieved from each individual query are given a ranking. This initial ranking is a combination of the score returned by the database query against the text index and the static relevance of the metadata element (within which the individual query applies). For instance, a restriction verified for the *title* is more important than a restriction verified in an element that contains *supplemental information.*
   b) Then, these lists of initial results must be merged according to the logic operators (*and, or*). This merging process will also update the ranking of each record depending on the logic operator. For instance, in the case of an *or* operator with two restriction operands, the ranking of a record verifying both the left and right restrictions will be updated with the sum of the initial ranking given by left and right restrictions.
   c) Finally, after the merging of results, they will sorted by descending ranking.

4. *Aggregation of results.*
   a) Once the initial list of records verifying the user restrictions is obtained, they will be collapsed into an aggregated list of results. See later the details about the implementation of this method.
   b) The ranking of the aggregated results must be also updated. For instance, a collection record retrieved because two child records verify the user restriction will be given the sum of children's ranking.
   c) Finally, the aggregated list of results must be sorted again.

5. *Metadata load.* Last, the user will require the download of the XML metadata contained in each metadata record of the final list of results. This XML data is not loaded in previous steps so as not to overload memory capacity. Furthermore, the catalog discovery services only allow the download of a selected fragment of results.

After describing the context of query answering, we will detail now the different strategies that have been considered for the granularity of the records that will be returned as answer to the user query. For a better understanding of these possibilities, figure 2.22 displays an example of the metadata records contained in a catalog, where the aggregation relations established among records are represented by means of hierarchical trees. The records with shaded background represent the records that verify the user query restriction before performing the aggregation of results.

1. *No aggregation.* This is not really an aggregation strategy because it does not apply any processing. However, sometimes one might be interested in avoiding the automatic aggregation of results. As regards the example, all the records with shaded background would be returned: $d, e, c, i, h, k, m, n, r, s$.

2. *Closest ascendant.* The idea is that if two or more records have an ascendant in the aggregation hierarchy, they will be minimized by returning the closest ascendant of these initial records. This method tries to return the most specific information but assuring that there are not two records of the same collection. For instance, in the example of figure 2.22 this method should return the records: a, f, k, m, p.

3. *Upper-level.* The idea of this strategy is to return uniquely records describing upper-level resources. If a record verifying the restriction belongs to a series of recursive collections, this method will only return the record that describes the upper-level ascendant. The objective in the example would be to return: a, f, j, m, o.

4. *Closest ascendant and depth filtering.* This is the same strategy as the "*closest ascendant*" but adding the restriction that only the records over a given depth in the aggregation hierarchy will be considered. This strategy uses a depth parameter greater than 0 (value 1 represents an upper-level resource, i.e. a record without parents). For instance, this strategy with level value of 1 would return uniquely the record m; a value of 2 would return c, h, k, m; and a value of 3 (the maximum depth) would return the same records as the "*closest ascendant*" strategy.

5. *Upper-level and depth filtering.* This is the same strategy as the "*upper-level*" but adding the restriction that only the records over a given depth in the aggregation hierarchy will be considered. Similar to "*Closest ascendant and depth filtering*" strategy, it requires a depth parameter with identical interpretation. As regards the example, this strategy would return uniquely the record m for level 1; a value of 2 would return a, f, j, m; and a value of 3 (the maximum depth) would return the same records as the "*upper-level*" strategy.

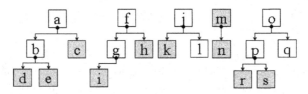

Fig. 2.22. A hierarchical set of records (shaded records verify user restrictions)

Although the catalog client should elect the most appropriate possibility for a particular search context, it is believed that the *closest ascendant* strategy facilitates a more general minimization, which is suitable for most scenarios. It provides the maximum detail but without including records that belong to the same collection.

With respect to the implementation of these strategies, it must be taken into account that intelligent query answering is closely related with the concept of Knowledge Discovery in Databases. Knowledge Discovery in Databases (also called data mining) refers to the nontrivial extraction of implicit, previously unknown, and potentially useful information from the data in databases (Frawley et al., 1991). Besides, as mentioned in section 2.4.2 our XML metadata is stored within a relational database with XML support. Thus, we studied the possible use of techniques for knowledge discovery in databases (Han et al., 1996).

Among these techniques, it is common the use of deductive databases or at least the use of deduction rules. As mentioned in (Ramamohanarao and Harland, 1994), deductive databases not only store explicit information in the manner of a relational database, but they also store rules that enable inferences based on the stored data to be made. The deductive database field has close links with the logic programming community, and much of the development of deductive database systems has centered around languages on Horn clauses (a syntax for expressing first order logic predicates) (Tärnlund, 1977; Ullman, 1989).

A Horn clause is generally written as

$$p(\tilde{t}) : - q_1(\tilde{t_1}), \ldots, q_n(\tilde{t_n})$$

where p and $q_1, \ldots, q_n$ are predicate letters, $n \geq 0$, and $\tilde{t}, \tilde{t_1}, \ldots, \tilde{t_n}$ are arbitrary (first order) terms, which may contain constants, variables and/or function symbols. All variables that occur in the terms are considered universally quantified at the front of the clause. The atom $p(\tilde{t})$ is referred to as the head of the clause, and $q_1(\tilde{t_1}), \ldots, q_n(\tilde{t_n})$ as the body of the clause. When n is 0, we refer to the clause as a *fact*. Otherwise, we refer to the clause as a *rule*. The semantics of these clauses can be interpreted as follows: the head is true if and only if all the atoms in the body are true. Finally, it must be mentioned that the Horn clauses language is usually extended so that the body of a clause is a conjunction of literals (i.e., an atom or the negation of an atom, rather than a conjunction of atoms alone).

In the deductive database field, all this logic is mapped into two main components of these databases: the Extensional Database (EDB) component consisting of facts, i.e. the tuples stored in relations; and the Intensional Database (IDB) consisting of rules, i.e. defined relations that do not exist in the database.

Following the deductive approach, we tried to model our aggregation strategies by means of Horn clauses. Figure 2.23 displays the deduction rules for each strategy. According to these deduction rules, deductive databases could infer which results are the final aggregated results to the user queries. For instance, figure 2.24 shows a refutation graph [11] demonstrating that record labeled with f in figure 2.22 is a final result according to the *closest ascendant* strategy.

---

[11] Refutation is a demonstration technique used in logic which consists in demonstrating that the negation of the initial objective is unsatisfiable.

No aggregation:
$verify(X, restriction) : - X$ complies with the 'restriction'
$finalResult(X, restriction) : -verify(X, restriction)$

Closest ascendant:
$whole(X, Y) : -$
$closestAscendant(X, Y, Z) : -$
        Z is the closest ascendant in the aggregation hierarchy of X and Y
$verify(X, restriction) : - X$ complies with the 'restriction'
$verify(Z, restriction) : - verify(X, restriction), verify(Y, restriction)$
        $, closestAscendant(X, Y, Z)$
$ascendantsNotVerify(X, restriction) : - whole(X, null)$
$ascendantsNotVerify(X, restriction) : - whole(X, Y), \neg verify(Y, restriction)$
        $, ascendantsNotVerify(Y, restriction)$
$finalResult(X, restriction) : - verify(X, restriction), ascendantsNotVerify(X, restriction)$

Upper level:
$whole(X, Y) : -$
$verify(X, restriction) : - X$ complies with the 'restriction'
$verify(Y, restriction) : - verify(X, restriction), whole(X, Y), Y \neq null$
$finalResult(X, restriction) : - verify(X, restriction), whole(X, null)$

Closest ascendant and depth filtering:
$whole(X, Y) : -$
$closestAscendant(X, Y, Z) : -$
        Z is the closest ascendant in the aggregation hierarchy of X and Y
$verify(X, restriction) : -$
        X complies with the 'restriction', depth of X greater than a given 'depth'
$verify(Z, restriction) : - verify(X, restriction), verify(Y, restriction)$
        $, closestAscendant(X, Y, Z)$
$ascendantsNotVerify(X, restriction) : - whole(X, null)$
$ascendantsNotVerify(X, restriction) : - whole(X, Y), \neg verify(Y, restriction)$
        $, ascendantsNotVerify(Y, restriction)$
$finalResult(X, restriction) : - verify(X, restriction), ascendantsNotVerify(X, restriction)$

Upper level and depth filtering:
$whole(X, Y) : -$
$verify(X, restriction) : - X$ complies with the 'restriction', depth of X greater than a given 'depth'
$verify(Y, restriction) : - verify(X, restriction), whole(X, Y), Y \neq null$
$finalResult(X, restriction) : - verify(X, restriction), whole(X, null)$

Notes:

1. Variables X, Y, Z represent metadata records.
2. $whole(X, Y)$ represents a whole-part relationship, being Y the parent record.
3. $whole(X, null)$ denotes a record without parent.

Fig. 2.23. Deduction rules

There are several deductive database systems that enable the interoperation with Oracle or other commercial databases for persistent storage. XSB[12] is an example of such a system. It is a Logic Programming and Deductive Database system which offers an interface to Oracle. This interface generates SQL code for Prolog queries on-demand, and translates Datalog clauses into SQL. Datalog clauses are Horn clauses where terms are only variables or constants (Ramamohanarao and Harland, 1994). However, the generated SQL is thought to perform queries against typical relations (tables), not to generate

---

[12] http://xsb.sourceforge.net/

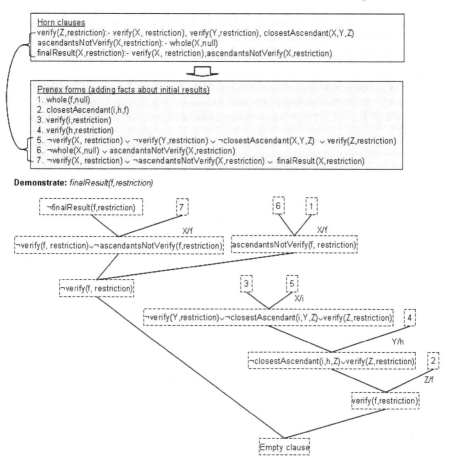

Fig. 2.24. Example of refutation graph according to *closest ascendant* strategy

text queries on large text columns. Therefore, the adoption of XSB seemed to be not feasible.

On the other hand, there are nowadays some projects that have tried to translate deductive database concepts to XML. For instance, the LoPiX (Logic Programming in XML) project (May, 2004) is the continuation and migration of the F-Logic/Florid project to XML. The LoPiX system is an implementation of XPathLog, a logic-based language for manipulating, querying and integrating XML data. Although this approach seems very attractive, it is still a research project (not commercialized) that lacks for the robustness of relational database storage.

After this initial study in the deductive database field, we considered that the direct adoption of an existing deductive system would involve high costs of integration, which would even imply the reconsideration of the initial hypothesis of storing XML data in a DBMS like Oracle. Furthermore, the deduction rules presented in figure 2.23 are located at a very specific place in the model. These rules only imply the navigation through the whole-part hierarchy that connects the metadata records. Other deductions, such as the automatic metadata inference through the aggregation relationship, have been already handled by other mechanisms presented in section 2.4.3. Therefore, we have finally opted for an ad-hoc implementation of these deductive mechanisms.

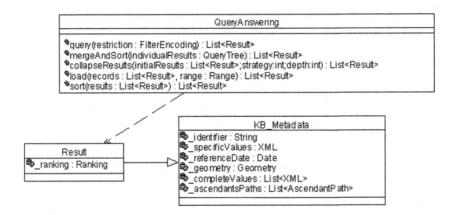

Fig. 2.25. Classes involved in query answering

Figure 2.25 shows the classes involved in query answering. The *QueryAnswering* class offers the interface methods to perform the tasks displayed in figure 2.21. In particular, the method *collapseResults* is the one in charge of the aggregation of results. The implementation of the *collapseResults* method is based on the comparison of the paths of ascendants (starting from the root) in the aggregation hierarchy of the results that verify the restriction imposed by the user. It receives a list of *Result* objects, which represent the initial list of results retrieved from the database as response of the user restriction. The *Result* class inherits from *KB_Metadata* and adds a special attribute to store the ranking of the result (attribute _ranking). But for the sake of efficiency, the *Result* objects received by the *collapseResults* method do not contain the XML metadata yet. Apart from the ranking, the only valid attributes are those providing identification, i.e. the _identifier and the _ascendantPaths attributes. This last attribute contains the paths of ascendants starting from the records at the top of the aggregation hierarchy. It must be mentioned that although these paths can be computed dynamically, they are pre-computed and stored at the database to speed-up the aggregation of results.

Figure 2.26 shows how the *collapseResults* method performs this minimization of results. In the worst case, the *upper-level* and *closest ascendant*

```
 1:  public static List collapseResults(List initialResults, int strategy, int depth){
 2:     // no processing is needed for NO_AGGREGATION strategy
 3:     if (strategy==NO_AGGREGATION) return initialResults;
 4:
 5:     // filter by depth if it is applicable
 6:     if ((strategy==CLOSEST_ASCENDANT_DEPTH_FILTERING)||
 7:         (strategy==UPPER_LEVEL_DEPTH_FILTERING))
 8:       initialResults = depthFiltering(initialResults,depth);
 9:     // collapse results for strategies: CLOSEST_ASCENDANT, UPPER_LEVEL
10:     // CLOSEST_ASCENDANT_DEPTH_FILTERING and UPPER_LEVEL_DEPTH_FILTERING
11:     List finalResults = new LinkedList(); // final list of collapsed results
12:     while (!initialResults.isEmpty()) {
13:       // extract first element of initialResults, which is stored in 'a'
14:       Result a = (Result) initialResults.remove(0);
15:       // obtain the root if the strategy is UPPER_LEVEL
16:       if ((strategy==UPPER_LEVEL)||(strategy==UPPER_LEVEL_DEPTH_FILTERING))
17:         a= a.getRoot();
18:       // compare 'a' with the rest of elements that remain in 'initialResults'
19:       ListIterator it = initialResults.listIterator();
20:       while (it.hasNext()) {
21:         Result b = (Result) it.next();
22:         // compare 'a' and 'b' to detect possible intersections
23:         Result intersection = a.getIntersection(b);
24:         // 'intersection' stores the common prefix of the descendant paths of 'a'
25:         // and 'b'. The ranking of 'intersection' is the sum of the ranking
26:         // of 'a' and 'b'
27:         if (intersection!=null) {// There is an intersection
28:           a=intersection;// 'a' is updated with the intersection between 'a' and 'b'
29:           it.remove(); // element 'b' is removed from 'initialResults'
30:         }
31:       }
32:       // append 'a' to the list of 'finalResults'
33:       finalResults.add(a);
34:     }
35:     // Sort the results by descending ranking
36:     sort(finalResults);
37:     return finalResults;
38:  }
```

Fig. 2.26. The *collapseResults* method

strategies, this algorithm much check all possible combinations of a pair of records (the number of combinations is $\binom{n}{2}$, involving a complexity in time of $O(n^2)$). If two records intersect in their paths, they are minimized with the appropriate record, the upper ascendant (*upper-level* strategy) or the closest ascendant (*closest-ascendant* strategy) found in the paths of ascendants. It must be remarked that this algorithm also updates the ranking of collapsed results. The *getIntersection* method invoked in line 23 returns a new *Result* (when a and b intersect) whose ranking is the sum of the rankings of a and b. Line 36 sorts the final list of results according to the descending ranking of the collapsed results.

In order to illustrate this implementation, we will show the trace of the *closest ascendant* strategy for the example of figure 2.22. In this case, the initial results would be represented by $(a, b, d), (a, b, e), (a, c), (f, g, i), (f, h), (j, k)$, $(m), (m, n), (o, p, r), (o, p, s)$. And the minimizations would be: $(a, b, d), (a, b, e)$ and (a, c) intersect in (a); (f, g, i) and (f, h) intersect in (f); (m) and (m, n)

It	a	b	initialResults	finalResults
1	(a,b,d)	(a,b,e)	(a,b,e),(a,c),(f,g,i),(f,h),(j,k),(m),(m,n),(o,p,r),(o,p,s)	
2	(a,b)	(a,c)	(a,c),(f,g,i),(f,h),(j,k),(m),(m,n),(o,p,r),(o,p,s)	
3	(a)	(f,g,i)	(f,g,i),(f,h),(j,k),(m),(m,n),(o,p,r),(o,p,s)	
4	(a)	(f,h)	(f,g,i),(f,h),(j,k),(m),(m,n),(o,p,r),(o,p,s)	
5	(a)	(j,k)	(f,g,i),(f,h),(j,k),(m),(m,n),(o,p,r),(o,p,s)	
6	(a)	(m)	(f,g,i),(f,h),(j,k),(m),(m,n),(o,p,r),(o,p,s)	
7	(a)	(m,n)	(f,g,i),(f,h),(j,k),(m),(m,n),(o,p,r),(o,p,s)	
8	(a)	(o,p,r)	(f,g,i),(f,h),(j,k),(m),(m,n),(o,p,r),(o,p,s)	
9	(a)	(o,p,s)	(f,g,i),(f,h),(j,k),(m),(m,n),(o,p,r),(o,p,s)	
10	(f,g,i)	(f,h)	(f,h),(j,k),(m),(m,n),(o,p,r),(o,p,s)	(a)
11	(f)	(j,k)	(j,k),(m),(m,n),(o,p,r),(o,p,s)	(a)
12	(f)	(m)	(j,k),(m),(m,n),(o,p,r),(o,p,s)	(a)
13	(f)	(m,n)	(j,k),(m),(m,n),(o,p,r),(o,p,s)	(a)
14	(f)	(o,p,r)	(j,k),(m),(m,n),(o,p,r),(o,p,s)	(a)
15	(f)	(o,p,s)	(j,k),(m),(m,n),(o,p,r),(o,p,s)	(a)
16	(j,k)	(m)	(m),(m,n),(o,p,r),(o,p,s)	(a),(f)
17	(j,k)	(m,n)	(m),(m,n),(o,p,r),(o,p,s)	(a),(f)
18	(j,k)	(o,p,r)	(m),(m,n),(o,p,r),(o,p,s)	(a),(f)
19	(j,k)	(o,p,s)	(m),(m,n),(o,p,r),(o,p,s)	(a),(f)
20	(m)	(m,n)	(m,n),(o,p,r),(o,p,s)	(a),(f),(j,k)
21	(m)	(o,p,r)	(o,p,r),(o,p,s)	(a),(f),(j,k)
22	(m)	(o,p,s)	(o,p,r),(o,p,s)	(a),(f),(j,k)
23	(o,p,r)	(o,p,s)	(o,p,s)	(a),(f),(j,k),(m)
Sorted final results				(a),(f),(m),(o,p),(j,k)

Table 2.5. Trace of the execution of *collapseResults* method

intersect in (m); and, (o,p,r) and (o,p,s) intersect in (o,p). Table 2.5 shows a trace of the status of variables in each iteration of the previous algorithm before executing sentence in line 23. As regards the ranking of the collapsed results, if we suppose that initially all results have a ranking of 1, last row of table 2.5 shows the sorted list of final results: (a) with ranking 3; (f), (m) and (o,p) with ranking 2; and (j,k) with ranking 1.

2.5 Building aggregation relations

The process of cataloguing a collection scenario is composed of three main steps:

1. The analysis of the structure of the collection. This consists of: detecting the metadata standard that will be used for describing components and the collection itself; finding the hierarchical structure of the collection and possible associated patterns; and establishing the inference mechanisms that the aggregation relationship should provide.
2. Design of metadata and aggregation relation types. If it were necessary, we should create the instances of *KB_MetadataType* and *KB_AggregationRelationType* to reflect the collection scenario that has been analyzed in the first step.
3. And finally the cataloguing of the components of the collection. Here, we should create: the instance of *KB_Metadata* that describes the collection; the instance of *KB_Aggregation Relation* that points at the appropriate

KB_AggregationRelationType and contains the description of the desired pattern; and then, the instances of *KB_Metadata* that describe the specific features of the components in the collection.

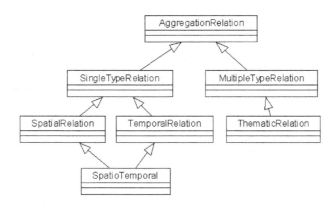

Fig. 2.27. Prototypical aggregation relations

Although this process may still seem arduous, it is alleviated by the fact that we could identify a small number of prototypical types of aggregations that may cover 90% of collections and independently of the metadata standard used. Figure 2.27 shows a hierarchy of these prototypical relations, which can be stored as instances of *KB_AggregationRelationType*. Apart from providing a conceptual distinction of aggregation relations, these predefined aggregation relations facilitate a knowledge (behavior stored in the attribute values) that may be reused in similar scenarios. Thus, steps 1 and 2 from previous process is usually reduced to the election of the prototypical aggregation relation type that applies in such scenario.

Now, some details about this hierarchy of aggregation relation types will be introduced. The names of the elements of the metadata standard ISO 19115 have been used to illustrate the features of these aggregation relation types. Nevertheless, these types are also applicable to the majority of metadata standards.

The *AggregationRelation* type is the parent type of the hierarchy. It represents a default aggregation relation and it offers two default aggregated functions to obtain the value of the *geographic location* (represented by means of a *bounding box*) and the *temporal extent* of the *whole* record. These functions, which are stored in the *_wholeInferredValuesSpecification* attribute, specify how to compute the minimum *bounding box* (*whole.boundingBox* = $\bigcup_{part}$ *part.boundingBox*) and *temporal extent* (*whole.temporalExtent* = $\bigcup_{part}$ *part.temporalExtent*) covered by the components of the collection.

The *SingleTypeRelation* type represents aggregations of resources which are obtained from the same source. They are equivalent to the single-type collections concept presented in (FGDC, 2002) or the *Portion/Mass* aggregation mentioned in (Winston et al., 1987) (see section 2.2.3). Inside this category, it is possible to distinguish *SpatialRelations, TemporalRelations and SpatioTemporalRelations*, which are detailed below.

The *SpatialRelation* type represents aggregations of resources which are spatially distributed, e.g. a collection that aggregates a set of components whose geographic extension correspond to the extension of tiles in a pre-established spatial division. As a special characteristic, the *_constraints* attribute of this relation type specifies that the *geographic location* element must be not null (otherwise we would not have a spatial collection), and that all the components should share the same *coordinate reference system* (at least projection system) and *spatial resolution*. Additionally, it is recommended to specify an appropriate *_pattern* in the corresponding instance of *KB_AggregationRelation*. These patterns are usually well known grids at different scales that are reused in multiple scenarios. They facilitate the creation and supervision of component metadata records because the cells (tiles) of these grids provide the *geographic location* information and probably other additional information (e.g., toponyms). Furthermore, these patterns can be used by catalog clients to facilitate map-based queries over the components of the collections.

The *TemporalRelation* type represents the aggregation of resources with similar characteristics (same source and geographic area) but taken at different instant times. This type of relation could be also assimilated to the *Member/Collection* aggregation (e.g., "a tree is part of a forest") mentioned in (Winston et al., 1987). As a special characteristic, the *_constrains* attribute specifies that: the *temporal extent* element must be not null (otherwise we would not have a spatial collection) and that all the components should share the same *coordinate reference system* (each component is an occurrence of the same area at a different moment) and *spatial resolution*. Additionally, it is recommended to specify an appropriate *_pattern* in the corresponding instance of *KB_AggregationRelation*, e.g. specifying the frequency for the creation of components in the collection.

The *SpatioTemporalRelation* type represents aggregations of resources that are spatially as well as temporally distributed. Usually, they follow patterns for both spatial and temporal distribution. The *_constraints* attribute of this relation type inherits the constraints already specified for *SpatialRelation* and *TemporalRelation* types.

The *MultipleTypeRelation* type represents aggregations of a wide range of geographic resources, probably originated by different sources, in order to perform a GIS study or project. Here, it is not easy to find a prototypical template of characterization. This type of relation is equivalent to the multiple-type collection concept presented in (FGDC, 2002) or the *Component/Integral Objects* aggregation mentioned in (Winston et al., 1987). Inside this category, it is also possible to distinguish a subtype called *ThematicRelation*.

The *ThematicRelation* type represents an aggregation of resources, each of them dealing with a different theme/subject. These subjects should belong to a pre-established thesaurus or controlled list of subjects. Thus the pattern attribute of the corresponding instance of *KB_AggregationRelation* should describe such controlled list or thesaurus. As a special characteristic, the _constraints attribute of this *KB_AggregationRelationType* specifies that at least all the components of the collection should have an overlapping *geographic location*; otherwise it is not possible to combine the information of the different layers. Additionally, it should be recommendable that all the layers share the same *coordinate reference system* and *spatial resolution*. But these last heterogeneities could be overcome by some kind of preprocessing.

2.6 Conclusions and future work

This chapter has presented a solution for the management of nested collections of resources. A general characteristic of the components of a collection is that they share a high percentage of meta-information. In order to facilitate this collection management, this chapter proposes the design of catalog system that is based on the use of a Metadata Knowledge Base component. The main features of this knowledge base component are the use of XML technologies and the improvement in the expressive power of the aggregation relations that define the components of a collection. This expressive characterization of aggregation relations facilitates the automatic inference of meta-information for both components and collections metadata records. Thanks to this metadata inference, it is possible to segregate the meta-information at the appropriate level of commonality or specificity, thus avoiding redundancy of information. However, this does not hinder applying the reverse process to obtain automatically complete descriptions of collections and components. Look at appendix A.1 for a discussion of the consistency of this compacted description of collections and components.

Although the concepts presented in this chapter are extensible to any type of digital library collection, the context of geographic information has been used to illustrate the solution presented here. In this context, the management of collections and series of resources is an important need and the knowledge representation model presented in this chapter may provide great benefits for the construction of metadata cataloguing systems integrated within geolibraries or Spatial Data Infrastructures.

First of all, this system will avoid the redundancy in metadata creation; metadata are only maintained in one place and inherited whenever it is needed.

Secondly, this system will facilitate the supervision of the metadata creation process. This system will enable the specification of patterns that the components of the collection should follow and thus, the status of cataloguing will be supervised by comparison with the patterns. For instance, in the case of a spatial collection, this system will be able to overlap the spatial pattern grid (the division of tiles for a specific scale) and the layer formed by the bounding boxes of the components already catalogued. Additionally, the metadata for the components of a spatial component could be graphically edited and facilitated by this spatial pattern (in the form of a coverage).

Another benefit of this system will be the possibility of providing discovery and presentation of metadata records at an aggregated or disaggregated level on user demand. The knowledge base can deduce whether an initial set of metadata results are describing components of the same collection, i.e. the knowledge base could find the metadata record that subsumes the initial results in the ascending whole-part hierarchy. Thanks to this, the system can present only an aggregated view of query results to the user in a first step, and a detailed view of the components metadata in a second step. Furthermore, for this second filtering the user can make profit of the collection pattern that defines the distribution of components.

A last benefit of the unified description of collections and components is that it can also help to generalize software for access and visualization of aggregated resources. For instance, an enhanced implementation of Web Map Servers (Beaujardière, 2002) could make profit of this cataloguing system to display automatically aggregations of datasets.

Finally, with respect to the future lines of the catalog system design proposed in this chapter, next steps should be oriented to give support for other types of relations that may be established between metadata records. Apart from the aggregation relation, other types of relations could be also benefited from the advantages that the metadata knowledge base approach provides: automatic metadata inference mechanisms, generation of statistics, navigation through relationships, and so on. In this sense, we have already detected a series of relations in the context of geographic information, which are detailed below.

One of these relations could be identified as a *version* relation. This relation reflects the association established between a set of source datasets and a dataset that has been derived from these source datasets. The semantics under this relation are similar to: the Dublin Core refinements *isVersionOf* and *hasVersion*; the *elaboration* relation defined by (Sathi et al., 1985); or the *MD_AggregateInformation* entity of ISO 19115 when the *associationType* attribute contains the value *source*. This relation could be even specialized depending on the type of transformation that is performed over the source datasets. Some examples of these specializations are the following:

- Coordinate Projection Transformations. Usually, data providers are required to produce and distribute their resources in different spatial reference systems (e.g., geographical coordinates, projected coordinates in UTM or Lambert). With the exception of the description of the *coordi-*

nate reference system and the addition of a *process step* in the *data quality* section, the rest of metadata describing the original resource created by the data provider or the derived resource is identical.

- Spatial Representation Transformation. Other times, data providers are required to convert the spatial representation of their resources. For instance, the geometry of a *mining quadrangle* feature may be represented alternatively with lines or intersection points. The only difference between the original and the derived metadata record is the description of the *geometric object* used (*type* and *count*) and the inclusion of a new *process step* in the *data quality* section.

- Operations on themes. (Rigaux et al., 2002) defines themes as the geospatial information (geographic location + attribute information) that correspond to a topic. Besides they define a series of operations (a theme algebra) that take one or more themes as input and return a theme. The metadata record describing the output theme can be almost automatically obtained in parallel to these operations. Some of these operations are the following: theme projection ($inputTheme, attribute_1, \ldots attribute_n \rightarrow outputTheme$) which consists in producing a new theme with a subset of the original descriptive information (a subset of the original attributes); theme selection ($inputTheme, attCond_1 \ldots attCond_n \rightarrow outputTheme$) which returns a subset of the geographic objects contained in a theme depending on some attribute conditions; theme union ($inputTheme_1, \ldots inputTheme_n \rightarrow outputTheme$) which consists in performing the union of sets of geographic objects having the same schema; theme overlay ($inputTheme_1, inputTheme_2 \rightarrow outputTheme$) returns a new theme whose geometry is the intersection of the input geographic objects and whose description is a combination of the participating descriptions; or theme merging ($inputTheme, condition \rightarrow outputTheme$) that performs the geometric union of the spatial parts of n geographic objects that belong to the same theme, under a condition supplied by the end user (e.g., it is usual to apply aggregate functions to a base data in order to obtain statistics/summaries, which remark features not perceived in the original (Fredikson et al., 1999)).

Other important type of relations could be entitled as a *revision* relation. This relation reflects the association between a source dataset and the datasets derived from the previous one by correcting or revising some attributes values. The semantics under this relation are similar to: the Dublin Core refinements *isReplacedBy* and *replaces*; and the *revision* relation defined by (Sathi et al., 1985). The metadata record describing the new resource is practically identical to the original one except for: the inclusion of a new *process step* in the *data quality* section; the update of data quality reports; and the modification of the *temporal extent* and *publication date*.

Another type of relation, which must not confused with a *revision* relation, is the *format* relation. This relation reflects the association between a source dataset and the datasets derived from the previous one by delivering the same contents but in a different format. The semantics under this relation are similar to the Dublin Core refinements *isFormatOf* and *hasFormat*. Once again, the metadata record describing the new resource is practically identical to the original one except for: the inclusion of a new *process step* in the *data quality* section; the update of the information about the *format* in the *distribution information* section.

And last, we have also identified a special type of relations identified as *high-level aggregation* relations. Apart from giving support for collections where all the metadata records describing the components reside in a local catalog, we may encounter that geographic resources and their metadata are distributed at different nodes of a spatial data infrastructure. An example of this situation may be a natural risk management infrastructure that requires cross-border coordination. In such a scenario, resources that are produced and maintained by the different parties responsible of each subregion must be merged to facilitate a harmonized vision of the area in conflict. There are metadata records describing each individual resource but they are distributed across the different geographic data catalogs. Thus, it would necessary to support a metadata record at the central catalog of the infrastructure to provide a general description of the collection of resources distributed at the different nodes. This central record should point at the individual records in the distributed catalog that describe the individual resources. Such an aggregation relation between the central records and the distributed records is called a high-level aggregation. More details about the initial steps to support such an aggregation relation can be found in (Béjar et al., 2003b).

3

Interoperability between metadata standards

3.1 Introduction

The term "interoperability" is usually defined as "the ability of two or more systems or components to exchange information and to use the information that has been exchanged" (IEEE, 1990). Obviously, the main obstacle for the interoperation of systems is the heterogeneity in data and services managed by these systems (Visser et al., 1997). In order to determine whether systems are heterogeneous one can focus on different characteristics and this yields different types of heterogeneity and consequently different types of interoperability. A commonly made distinction is that between syntactic (solving syntactic heterogeneity) and semantic interoperability (solving semantic heterogeneity) (Kolodziej, 2003). The syntactic interoperability is concerned with the technical level, i.e. it refers to the ability for a system or components of a system to provide information portability and interapplication as well as cooperative process control. It comprises intercommunication at communication level protocol, hardware, software, and data compatibility layers. The semantic interoperability, in contrast, deals with the domain knowledge necessary for informatics services to "understand" each other's intentions and capabilities. A more detailed categorization can be found in (Sheth, 1999), where four types of heterogeneity are distinguished: system heterogeneity (e.g., use of different operating systems and computing platforms), syntactic heterogeneity (e.g., differences in machine readable aspects of data representation), structural heterogeneity (e.g., schematic heterogeneity that particularly appears in structured databases), and semantic heterogeneity (equivalent to the semantic interoperability defined in (Kolodziej, 2003)). This second division is comparable with the first one because the first three types are instances of the syntactic interoperability defined in (Kolodziej, 2003).

The creation of standards and the existence of agreed conventions have facilitated enormously the syntactic interoperability. For instance, standards like CORBA (Orfali et al., 1999) facilitate the interoperation of systems which may have been implemented with different programming languages and in different computing platforms; HTML is a language for the creation and presentation of Web contents with an agreed syntax; or standards like UML (Booch et al., 1998) facilitate structural interoperability by enabling the definition commonly understood application schemas. However, the syntactic interoperability is not enough to understand data and services (Ostman et al., 2002). For instance, one may receive a file in a standardized format, e.g. a file in SHAPE file format (proprietary format used by ArcView GIS tool[1]) containing a set of polygons, but this does not informs about its content and use. At first glance, one can not distinguish whether these polygons represent lakes, nature reserves or provinces. Therefore, it results vital to improve the semantic interoperability.

The use of metadata describing data and services facilitates the semantic interoperability. Promoting a commonly understood set of descriptors, it increases the possibility of semantic interoperability across disciplines. For instance, networks of library catalogs, which use agreed metadata schemas like MARC (U.S. Library of Congress, 2004b), facilitate search and retrieval of data with a high degree of accuracy while resting assured of its potential use and authenticity. Nevertheless, one may also find heterogeneity in the schemas used for metadata. Networked knowledge organization systems typically contain objects which are described using a multitude of diverse metadata schemas (Hunter, 2001). Considering the Web as the biggest networked knowledge organization system example, one can figure out the semantic interoperability problems that this implies [2]. The use of disparate description models interfere with the ability of search engines to search across discipline boundaries. Hence machine understanding of metadata descriptions which conform to schemas from different domains is a fundamental requirement for access to information within networked knowledge organization systems. This chapter will be devoted to this problem of metadata interoperability.

Metadata descriptions from different domains are not semantically distinct but overlap and relate to each other in complex ways. As the number, size and complexity of the metadata standards grow, the task of facilitating metadata in different standards becomes more difficult and tedious. In order to minimize the cost of time for the creation and maintenance of metadata and to maximize its usefulness to the wider audience of users, it should be desirable to use a unique metadata standard in storage labours and provide automated views of metadata in other related standards. Furthermore, other times the metadata interoperability is not uniquely a cross-domain problem. Within the same

---

[1] http://www.esri.com/

[2] A good example to illustrate the diversity of metadata standards is the MetaMap project of the University of Montreal, available at http://www.mapageweb.umontreal.ca/turner/meta/english/. Taking the metaphorical form of a subway map, this project helps users navigate in the "metaspace" and it establishes the relationships among the processes of information management, the institutions with expertise in managing information, and the types of information files that are managed.

domain, a metadata describing an instance of an entity A can be derived from a set of metadata entries describing instances from an entity B. For instance, the bibliographic records describing a collection of books can be summarized to obtain the metadata which describes the entire collection. But once again, it should be desirable to maintain uniquely the source metadata entries and generate automatically the derived metadata.

The tendency of the current cataloguing systems is to interchange metadata in XML according to the specific standard required by each user on demand, that is to say, providing different views of the same metadata. In order to maintain this interoperability across related metadata standards, it is necessary the creation of software systems able "to speak several metadata dialects", that is to say, systems that provide crosswalks between metadata standards. According to the Dublin Core Metadata Glossary (DCMI, 2001): "A crosswalk is a table that maps the relationships and equivalencies between two or more metadata formats. Crosswalks or metadata mapping support the ability of search engines to search effectively across heterogeneous databases, i.e. crosswalks help promote interoperability".

Let us imagine a scenario where three different metadata-databases store meta-information that describes the elements from a library (books, reports and other kinds of documents), events (movies, theaters, recitals, etc) and geographic data (maps, satellite images, etc) respectively. These databases can be used for providing specialized high-level services such as tourist information (events and publications can be linked with data for traveling to a tourist destination) or cultural information (publications can be linked to an event, and it could be useful to have maps for accessing to the places where the event occurs). The problem is that the standard used in each metadata-database belongs to a distinct domain and it will be necessary to unify the metadata-access (search and retrieval) methods. Figure 3.1 displays the scenario described above and the different databases that must be integrated.

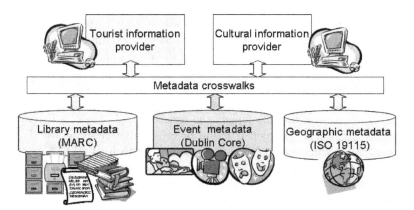

Fig. 3.1. Crosswalk use cases

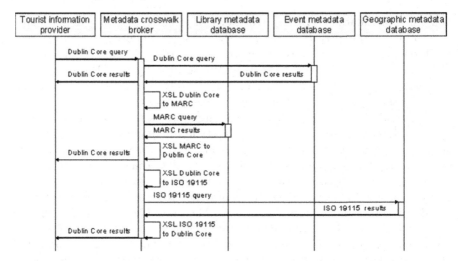

Fig. 3.2. Crosswalks applied for the tourist information provider use case

If we had to develop a system for the tourist information provider, this system should use and homogenous mechanism for querying and accessing the three databases. That is to say, the metadata schema of the tourist information provider system should be independent of the metadata representation used by the three databases. For instance, in the referred example, the tourism information provider could query the system and managing the information using Dublin Core (DCMI, 2004; ISO, 2003d; ANSI, 2001), whereas the cultural information provider manages only MARC metadata (U.S. Library of Congress, 2004b). The aforementioned homogenous mechanism should be a crosswalk broker facilitating the integration and coordination of crosswalks when needed. This broker should consist of a repository of crosswalks (Dublin Core ↔ MARC, Dublin Core ↔ ISO 19115 (ISO, 2003a), and MARC ↔ ISO 19115 in the previous example) and the software for activating and processing these crosswalks when needed. Figure 3.2 shows an example of the sequence of crosswalks applied in order to query the databases and obtain the results [3].

Presumably, given that de-facto standard for the exchange of metadata is XML, final implementation of crosswalks should be based on XSL technology (W3C, 2004a) (figure 3.2 depicts the order of appliance of different XSL stylesheets). However, the construction of crosswalks between standards is much more than the use of a series of programming technologies. A crosswalk specifies the mapping between two related standards, thus enabling communities that use one standard to access the content of elements defined in another

---

[3] More details about this use case and the applicability of crosswalks in multidisciplinary scenarios can be found in (Zarazaga et al., 2003).

one. Unfortunately, the construction of crosswalks constitutes a difficult and error-prone task that requires deep knowledge and vast experience with the standards. The knowledge required to construct a crosswalk is particularly problematic since each metadata standard has been developed frequently in an independent form and therefore different terminology, specialized methods and processes are used. Moreover, the maintenance of crosswalks between metadata standards which are not stable and subject to changes is even more problematic due to the additional requirement of adjusting crosswalks to historical versions. For that reason, the harmonization in the consistent specification of related metadata standards is vital to the development of crosswalks. Thanks to this harmonized specification, it is easier to match the metadata elements of the different standards. The objective of this chapter is to present the process followed to carry out a series of crosswalks that enable interoperation across some of the most relevant standards for geographic information metadata.

Issues arisen in the development of metadata crosswalks are not constrained to a specific application domain. That is to say, similar problems must be solved for digital libraries metadata or for more specific geographic information metadata. The rest of this chapter is structured as follows. Next section presents the work related with the metadata interoperability with special interest in the geographic information domain. Then, a general process will be proposed to formalize metadata standards and construct crosswalks. Next, the experience in developing several crosswalks using this process is explained. Finally, this chapter ends with a section of conclusions and future work.

3.2 Related work

According to (Hunter, 2001), there are three main scenarios in which interoperability among metadata descriptions is required: to enable a single search interface across heterogeneous metadata descriptions; to enable the integration or merging of descriptions which are based on complementary but possibly overlapping metadata schemas or standards; and to enable different views of the one underlying and complete metadata description, depending on the user interests, perspective or requirements. The solutions to handle the problem of metadata interoperability in some or all of these scenarios may be classified into two main approaches: solutions that are based on the use of ontologies (i.e. establishing or inferring relationships between the metadata vocabularies employed by the different metadata standards); and the creation of specific crosswalks for one-to-one mapping. Next subsections will present the work related with these approaches.

3.2.1 Ontology based semantic interoperability

The impact of the Internet as the biggest platform for the distribution of resources has motivated the birth of a great deal of initiatives that aim at solving the problem of semantic interoperability on the Web. As mentioned in section 1.6, an ontology is defined as an explicit specification of some shared vocabulary or conceptualization of a specific subject matter, and it seems to be an adequate methodology that helps to define a common ground between different information communities. An example of an ontology-based solution for interoperability among distributed data repositories could be the OBSERVER system (Mena, 1998). This system provides an architecture for query processing in global information systems that supports interoperation across ontologies. In this system, each ontology defines the terms used to access the contents of a specific data repository, i.e. the ontology compiles the terms which are later mapped to the specific data structures (names of entities and attributes). And an inter-ontology contains the relationships relating the terms in the different ontologies, which enable the translation of the user query to the specific ontology of each distributed repository.

Nowadays, most of ontology-based approaches for semantic interoperability rely on RDF technologies as the basis for information sharing (Pundt and Bishr, 2002). Furthermore, these approaches are closely related to a new conception of the Web: the Semantic Web. According to (Berners-Lee et al., 2001; W3C, 2004b), "the Semantic Web is an extension of the current web in which information is given well-defined meaning, better enabling computers and people to work in cooperation". RDF (Resource Description Framework) (Manola and Miller, 2004) is a W3C recommendation for modeling and exchanging metadata. The major advantage of RDF is its flexibility. RDF is not really a metadata standard defining a series of elements. On the contrary, it can be considered as a meta-model that contains other metadata schemas or combinations of them. RDF uniquely defines a simple model for describing the interrelationships among resources in terms of named properties and values. But for the declaration and interpretation of those properties, a complementary technology of RDF is needed. This complementary technology is RDFS, which stands for RDF Schema although it has been recently renamed as RDF Vocabulary Description Language (Brickley and Guha, 2004). RDFS provides a rich set of constructs to define and constrain the interpretation of vocabularies used in a certain information community. In fact, a RDFS document defines the ontology that is used to construct particular RDF documents in an information community. That is to say, RDFS can be used to define the semantic meaning of metadata elements contained in a metadata standard or schema, viewing the structure of metadata schemas as ontologies. In this sense, an instance of RDFS could be seen as the ontology of metadata elements used for a particular profile. A more general solution for interoperability should be based on the use of ontologies that define the semantic meaning of metadata elements contained in each metadata standard or schema. Moreover, RDFS documents (defining ontologies) can reuse other ontologies that may be located and controlled in other places on the Internet. As a result, if different information communities define their domain ontologies by means of RDFS and publish their metadata in RDF, other information communities can check whether these metadata (including the semantics) is usable or not.

An example of this kind of approaches is the work presented in (Hunter, 2001). There, the ontology is implemented as a thesaurus, named MetaNet, whose objective is to provide the semantic knowledge required to enable machine understanding of equivalence and hierarchical relationships between metadata terms from different domains. The scope of this thesaurus is limited to the most significant metadata models/vocabularies used for describing attributes and events associated with resources and their life cycles. This encompasses metadata vocabularies from the bibliographic, museum, archival, record keeping and rights management communities. MetaNet has been developed by performing WordNet (an upper level ontology) searches of the core terms used in the different domains. In order to implement dynamically the interoperability, this work provides an RDFS representation of the MetaNet thesaurus together with an XML stylesheet. This stylesheet parses an input metadata description and searches the MetaNet RDFS representation for the elements in the output metadata standard that are equivalent to the input element names.

As alternatives to RDF technologies for resource description and knowledge representation on the Web, there are some proposals like SHOE language, which can be found in (Heflin and Hendler, 2000). This work remarks the fact that RDFS representation is more limited than most artificial intelligence ontologies because it does not possess any mechanisms for defining general axioms (rules that allow additional reasoning). On the contrary, SHOE is presented as an ontology-based knowledge representation language designed for the Web that permits the discovery of implicit knowledge through the use of taxonomies and inference rules. The syntax of this language is defined as an application of SGML that extends the HMTL DTD, primarily because XML was still evolving when SHOE was created. SHOE ontologies are made publicly available by locating them on web pages. Then, ordinary web pages (the resource itself) are extended with special tags to include instances of the entities defined by a referenced SHOE ontology. Finally, the interoperability in SHOE is through use of the ontology extension and renaming features (two categories are similar to the extent that they share the same supercategories).

More specifically within the context of geographic information, another example of ontology-based interoperability solution is the system presented in (Weißenberg and Gartmann, 2003). There, an ontology architecture is used to offer personalized Geo-Services to athletes, journalists and spectators in Olympia 2008. Different metadata standards are used to describe the different geo-services. These metadata standards (e.g. ISO 19115 and Dublin Core) are modeled as ontologies using F(rame)-Logic (Kifer et al., 1995) and semantic technologies are used to match these ontologies and enable semantic queries.

As it has been seen, these approaches offer flexible solutions for interoperability. However, this ambitious aim of flexibility may also imply a lack of accuracy in the mappings performed. The ontology based solutions presented until now do not consider the local structural constraints imposed by the different specific domains, e.g. parent/child relationships; cardinality/occurrence constraints; datatyping, enumeration and formatting constraints on the element values. The SHOE approach even defines its own metadata encoding language. As it is stated in (Hunter, 2001): "the wider the targeted scope of interoperability, the more difficult it is to achieve accurate, precise mappings". For a small set of metadata standards, whose syntax and semantics are relatively fixed and constrained, hardwired crosswalks establishing the mapping between metadata terms (from specific standards) may result more adequate than ontology-based solutions. That is precisely the case in the geographic information context. Furthermore, most of ontology-based solutions, e.g. OB-SERVER or the project for Olympia 2008, are only focused on providing a single search interface across heterogeneous metadata descriptions. But this chapter aims at giving a solution which enables different views of underlying complete metadata descriptions as well.

3.2.2 Crosswalk based semantic interoperability

There is a big experience in developing mappings among several standards and different domains. For instance, interesting collections of links to metadata-crosswalk initiatives can be found through the Web sites of the UK Office for Library and Information Networking [4] and the Metadata Architecture and Application Team of the National Digital Archives Program in Taiwan [5]. There, it is possible to find several mappings among the main metadata standards (specially those used for library metadata): from MARC standards to Dublin Core; from Dublin Core to EAD (Encoded Archival Description) (U.S. Library of Congress, 1998); from Dublin Core to GILS (a Z39.50 metadata profile for the U.S Government Information Locator Service); or from Dublin Core to GCMD DIF (Directory Interchange Format) (Vogel and Northcutt, 1999).

Within the context of metadata for museum content (Sherwood, 1998), the Canadian Heritage Information Network (CHIN)[6] compiles a list of crosswalks and related resources that may be of use to museums. Some representative examples of the resources referenced could be the "Crosswalk of Metadata Element Sets for Art, Architecture, and Cultural Heritage Information and Online Resources" or the "Mapping from CHIN Natural Sciences Data Dictionary to Darwin Core". The first example is a crosswalk developed by the Getty Research Institute and within the mapped standards it includes: the

---

[4] http://www.ukoln.ac.uk/metadata/interoperability/

[5] http://www.sinica.edu.tw/~metadata/tool/mapping-foreign.html

[6] http://www.chin.gc.ca/

Categories for the Description of Works of Art (created by the Getty Research Institute to describe art databases); the VRA Core Categories (created by the Visual Resources Association Data Standards Committee); Dublin Core; Object ID (an international standard for describing cultural objects); the CIMI Access Points (created by the Consortium for the Computer Interchange of Museum Information); or the Guide to the Description of Architectural Drawings (created by the Getty Research Institute). And the second example enables museums using the CHIN Natural Sciences Data Dictionary to transform metadata into the Darwin Core profile, a profile describing the minimum set of standards for search and retrieval of natural history collections and observation databases.

As far as geographic information metadata is concerned, it is worth mentioning the work done by the European projects MADAME (Methods for Access to Data and Metadata in Europe) and ETeMII (European Territorial Management Information Infrastructure). Within the documents delivered by these closely related projects (Craglia, 2001), two metadata crosswalks were proposed for convergence towards ISO 19115 and Dublin Core. On one hand, they defined a mapping between ISO 19115 (a draft version) and Dublin Core elements. And on the other hand, they also proposed a mapping for migrating metadata compliant with the European norm prENV 12657 (CEN, 1998) towards a draft version of ISO 19115.

As concerns the interoperability between the CSDGM and the ISO 19115 standards, the Canadian Geospatial Data Infrastructure has developed a crosswalk, which can be found in (GeoConnections, 2001; Teng, 2000). The discovery portal of this infrastructure [7] offers data products catalogued in accordance with the CSDGM standard but it plans to support the ISO 19115 standards in future versions.

Additionally, the DGIWG (Digital Geographic Information Working Group) Metadata Work Program, supported by NIMA (National Imagery and Mapping Agency of United States), offers a crosswalk between ISO 19115 and the CSDGM standard too. This program is taking a leading role in developing an implementation model and XML-Schema of the ISO 19115 metadata standard (officially known as ISO 19139 (ISO, 2003b)) and provides a Metadata Development Efforts Website [8] to coordinate the metadata standardization efforts of several organizations.

On the other hand, the own FGDC, in charge of defining the CSDGM standard, provides a mapping between CSDGM and Dublin Core [9]. Moreover, this institution offers a metadata parser tool (called *mp*) that is able to generate an HTML output where CSDGM elements are mapped to Dublin Core elements in the META tags of an HTML document. The intended use of

---

[7] http://geoconnections.ca

[8] http://metadata.dgiwg.org/

[9] The mapping is available at http://geology.usgs.gov/tools/metadata/tools/doc/dublin.html.

META tags, included in the HEAD section of an HTML document, is to advertise the content of a Web page, thus making this meta-information visible to search engines.

Other works like (Chandler et al., 2000) have also proposed the conversion of CSDGM towards more generic standards like MARC or Dublin Core. The motivation for this conversion was due to the unsuccessful results (on average) obtained from queries directed at nodes of the FGDC Clearinghouse [10]. Therefore, it was proposed to convert CSDGM metadata into more widely used metadata standards, and thus include the original metadata in systems other than the FGDC Clearinghouse. In particular, the objective of this work was to obtain a converter able to insert metadata into the Cooperative Online Research Catalog (CORC). CORC was an initiative sponsored by the Online Computer Library Center (OCLC)[11] that aimed at integrating Dublin Core and MARC21 metadata into a single system. And nowadays, this initiative has become a private online service called *Connexion*[12] , which enables the access to the OCLC WorldCat (a worldwide union catalog maintained collectively by more than 9,000 member institutions).

And most specifically within the context of environmental geographic information, it is worth mentioning the work done by two projects: EIONET (European Environment Information and Observation Network) [13] and UDK (Umwelt Data Katalog) [14]. On one hand, the EIONET is a European network for the diffusion of environmental data that has defined a mapping between ISO 19115 and the metadata used in the GELOS (Global Environmental Locator Service) service. And on the other hand, UDK proposes the mapping between ISO 19115 and the metadata used for the German environmental data catalog.

One thing in common from these existent works is that almost no-one offers details about the process followed to obtain the mappings. Two exceptions are probably the works presented in (Woodley, 2000) and (Pierre and LaPlant, 1998), which are more focused on presenting the problems in crosswalk creation than in delivering the results of a particular mapping. The first work presents some of the common misalignments in crosswalks creation. And the second one provides many of the key issues involved in crosswalk development and identifies those areas in which harmonization can contribute. Its main contribution is the delineation of the general issues involved in the harmonization of metadata standards and in the development of crosswalks between related metadata standards. Many concepts and ideas presented in it have been used as a base for the development of the work presented in this chapter.

---

[10] http://www.fgdc.gov/clearinghouse/clearinghouse.html

[11] http://www.oclc.org/

[12] http://www.oclc.org/connexion/

[13] http://eionet.eu.int/

[14] http://www.umweltdatenkatalog.de/

Finally, another conclusion is that most of these works do not include any other result apart from the table that maps the relationships and equivalencies among the standards. Very few of them offer a tool to perform the translation. However, once a high-level mapping has been obtained, it should be interesting to have a semi-automatic tool able to make the low-level translation. In this sense, (Popa et al., 2002; Fagin et al., 2003) present a semi-automatic tool called Clio that enables the mapping between any combination of XML and relational schemas, in which a high-level, user specified mapping is translated into semantically meaningful queries that transform source data into the target representation.

3.3 Construction of crosswalks between metadata standards

Metadata interoperability is a problem not very different from the interoperability of heterogeneous databases. Data exchange is the problem of taking data structured under a source schema and creating an instance of a target schema that reflects the source data as accurately as possible. Semantic heterogeneity in databases has been studied extensively in the database field. For instance, (Ceri and Widom, 1993) presents four categories of semantic conflicts: naming conflicts (different databases use different names to represent the same concepts); domain conflicts (different databases use different values to represent the same concept); meta-data conflicts (same concepts are represented at the schema level in one database and at the instance level in another database); and structural conflicts (different databases use different data organization to represent the same concept). Saving the distance, crosswalks aim at solving all these conflicts that also arise in the conversion between two metadata standards.

This section presents the steps of the process that has been followed to construct a series of crosswalks between standards and that simplifies its implementation by means of the use of formal specifications and automated mechanisms. The process has the following steps (see figure 3.3):

1. Harmonization: This phase aims at obtaining a formal and harmonized specification of both standards.
2. Semantic mapping: This phase establishes the mapping between the elements in the source standard and the elements in the target standard. Although it seems to be a simple task, it requires a deep knowledge of the origin and target standards. According to the categorization of (Ceri and Widom, 1993), this phase would solve the naming conflicts and it would detect the meta-data conflicts.
3. Additional rules for metadata conversion: Apart from the semantic mapping, it should be necessary to provide additional metadata conversion rules in order to solve problems such as different level of hierarchy, data

type conversions, etc. According to the categorization of (Ceri and Widom, 1993), this phase would be devoted to solve the domain and structural conflicts.

4. Mapping implementation: The last objective of the process is to obtain a completely automated crosswalk by means of the application of some type of tool. In this way, maintaining only one metadata standard, searches and views can be provided according to the different families of metadata standards.

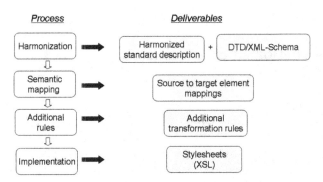

Fig. 3.3. Process steps

The following subsections present further details of each one of these steps.

3.3.1 Harmonization

Many of the metadata standards use similar properties in the definition of their content elements. Some examples of similar properties could be: a unique identifier for each metadata element (for example: tag, label, identifier); a semantic definition for each element; the mandatory, optional or conditional character of each element; the multiplicity or allowed number of occurrences of an element; the hierarchical organization with respect to the rest of elements; or constraints on the value of an element (e.g. free text, numerical range, dates or a predefined code list). If the way to express those properties were fixed, every metadata standard could be described in a similar way. Consequently, similar processes could be applied to related metadata standards, thus simplifying not only standards implementation but also the development of new crosswalks between them.

The generalization and formalization in the specification of metadata standard properties are usually done by means of a canonical representation or a specification language. This procedure is analogous to the specification of a programming language syntax using the well-known notation Backus-Naur-Form (BNF) (Backus et al., 1963). In fact, thanks to the circumstance that most standards use XML as exchange and presentation format, they also provide a DTD or XML-Schema that describes formally their syntax.

Nevertheless, a mere syntactic description of a metadata standard is not enough to store all the information necessary to automate the development of crosswalks. For instance, a minimum set of data types must be defined as a basis to obtain from it the derived data types that are required to represent all the elements in the target standard. And in addition to this, as it happens with BNF, a metadata specification does not contain information about the semantics of elements. Therefore, apart from the DTD or XML-Schema, we propose an extended and harmonized definition of each metadata element. In order to select the descriptors for describing consistently a metadata element we took into account: the standard ISO 11179-3 (ISO, 2003e), which forms part of the larger standard ISO 11179 ("Specification and Standardization of Metadata Elements") and specifies the basic attributes required to describe metadata items; and a guidelines document for the construction of Dublin Core application profiles (Baker et al., 2003), which has been issued by CEN (European Standardization Committee) and establishes a set of attributes to describe the elements included in the application profile. The descriptors that have been finally selected are:

- *Identifier*. This is the identifier given by the standard to identify the element. For instance, ISO 19115 uses the line number of the dictionary where these elements are defined. And other standards like Dublin Core define a URI (Universal Resource Identifier) for each element.
- *Name*. This is the long name (descriptive name) assigned by the standard to this element.
- *Obligation*. This descriptor indicates whether a metadata element shall always be present or sometimes be present (i.e., whether this element must contain a valid value). This descriptor may have the following values: M indicating that the element is mandatory and shall be present; C indicating that the element is conditional and it only shall be present under special conditions (see descriptor *condition*); and O indicating that the element is optional and may not be present.
- *Maximum Occurrence*. It describes any limit to the repeatability of the element. It may have the following values: 1 indicating that it has one value at maximum (it is non-repeatable); other values greater than 1; and N indicating that there is no limit (the element is repeatable).
- *Datatype*. This descriptor indicates the type of data that can be represented in the value of the element. Examples of datatypes are: 'character', 'ordinal number', 'integer' or 'character string'.
- *Definition*. It contains the description of an element that clearly distinguishes it from other metadata elements.
- *Comment*. It provides any additional information about the term or its application.

- *Condition.* It describes the condition or conditions according to which a value shall be present.
- *Path.* This descriptor contains the XPath (Clark and DeRose, 1999) expression that is needed to access the value of this element in the XML encoding of a metadata record. That is to say, it contains the sequence of XML tags that is necessary to browse until obtaining the value of an element. This is useful to facilitate the later implementation of crosswalks. Furthermore, these XPath expressions encode implicitly the nested structure of sections and subsections of the standard.

Finally, we propose in this step to create a database containing the harmonized definitions of the metadata elements. For instance, figure 3.4 displays the relational model of such database. Each element definition would correspond with a row of the relational table *ELEMENT*. Anyway, simpler tools as an Excel sheet could be used for this harmonized definition of elements. In this case, it would be recommendable that all these definitions (rows) should be sorted by the order of sections and subsections in the standard. Moreover, it should be desirable to indent the *name* of each element according to the hierarchical structure of the standard.

3.3.2 Semantic mapping

The most important task in the development of crosswalks is the one in charge of determining the semantic correspondence between the elements of the standards to be mapped (Pierre and LaPlant, 1998). This task implies the specification of a mapping between each element in the origin standard and the element that is semantically equivalent to this one in the target standard. For that purpose, it is very important to count on a clear and precise definition of each-standard elements.

Additionally, it is also frequent to find in this phase those conflicts classified as meta-data conflicts in (Ceri and Widom, 1993). These conflicts arise when the same concept is expressed at the schema level (i.e., the standard defines an explicit element for this concept) in the source standard and at the instance level (i.e. the concept is expressed as the value of an element) in the target standard. For instance, the CSDGM standard (FGDC, 1998) defines four different elements (*theme, place, temporal, stratum*) to classify the different types of keywords that may be included in the description of a geographic resource. However, other standards like ISO 19115 define a unique element (*descriptiveKeywords* association between the *MD_Identification* and the *MD_Keyword* classes) which contains a subelement (attribute *type* in *MD_Keyword*) whose value indicates the type of keyword. Therefore, the mapping between the four different elements of CSDGM and the unique element of ISO 19115 depends on the value given to the element specifying the type in ISO 19115.

With respect to the way of specifying these semantic mappings, many metadata standards already provide a semantic mapping with standards of related metadata; frequently this mapping appears in the form of a table in an annex of the standard. In the process that appears here and following the structure of the crosswalks database displayed in figure 3.4, we should fill the rows of the relational table *ASSOCIATION* that establish the link between a source element and a target element. It can be observed that this relational table includes a *COMMENT* attribute, which is oriented to clarify possible problems such those derived from the aforementioned meta-data conflicts. Finally, in case of using simpler tools like Excel, an additional sheet should be created in order to establish the relationship between the identifier of source elements and the identifier of target elements.

3.3.3 Additional rules for metadata conversion

A crosswalk is a set of transformations that applied to a set of elements in the source metadata standard produce, as a result, an equivalent content in the target standard, which has been properly modified and redistributed to meet the requirements of the analogous elements. Therefore, a completely specified crosswalk must consist of a table of the semantic mappings accompanied by a metadata conversion specification. This specification contains the additional transformations required to convert the metadata document whose contents fulfill the source standard into a document whose contents fulfill the target standard. Following subsections present the different metadata conversion problems that may arise.

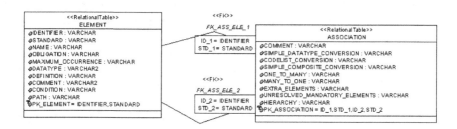

Fig. 3.4. Database of crosswalks

At the end of this phase and following the structure of the crosswalks database in figure 3.4, these rules would be stored in the different attributes of the relational table *ASSOCIATION*. And if it were necessary any additional information, annex documents should be created and attached to the crosswalk documentation.

Content Conversion

Frequently, metadata standards restrict the contents of each element to a particular data type, range of values or controlled vocabulary. In some cases, two analogous in elements in different standards may have different content restrictions. It has been identified that the most frequent cases are the following:

- Simple datatype conversions. By simple datatypes it is meant those common datatypes that are usually predefined in traditional programming languages such as numbers, characters, booleans, strings or dates. Thus, this category includes the conversions that are required for those datatypes between the source and the target standard. For instance, it could happen that a text value must be transformed into a numerical value or a date value.
- Code-lists conversions. Often, standards define elements whose values must be constrained to a controlled vocabulary, usually in the form of code-lists. In such cases, specific rules are required to establish the correspondence between the initial element whose values may be specified as free text and a target element whose value is constrained to a controlled vocabulary. Moreover, when mapping two elements restricted to different controlled vocabularies, it is necessary to establish the relationship between values on one-to-one basis.
- Composite to simple datatype conversions. Another typical case is the conversion between a simple datatype value and a composite datatype value. By composite datatypes it is meant those datatypes that would be equivalent to records in structured programming, which consist of two or more fields. For example, a crosswalk for Dublin Core to ISO 19115 standard should map the Dublin Core *creator* element to an instance of the *CI_ResponsibleParty* class (datatype), which consists of a large number of attributes (*individualName, organizationName, contactInfo,...*), some of them composite as well. In this case, it must be indicated how to extract correctly the content of the source element and map it to the corresponding attributes.

Each rule for the previous cases would be stored in the attributes *SIMPLE_DATA_TYPE, CODELIST_CONVERSION* and *SIMPLE_COMPOSITE* of the relational table *ASSOCIATION* (see figure 3.4).

Element to element mapping

One of the main problems that must be solved in one-to-one element mappings are those related with the *obligation* and *maximum occurrence* in each standard. The trivial case is the mapping between two elements that share identical properties, e.g. a mandatory non-repeatable element which matches with a mandatory non-repeatable element in target standard. However, for the rest of combinations the crosswalk must apply special rules, which could even imply the loss of information. These special cases can be classified in the following categories:

- One to many. In most cases, a one-to-many map is trivial; an occurrence of the source element maps to a single occurrence in the target element. However, there are cases where the mapping requires more explicit resolution. For example, the source standard may contain a non-repeatable "keywords" element and according to its definition the content of this element consists of one or more keyword values separated by commas. Nevertheless, this element should match with a repeatable element in the target standard, that is to say, an occurrence for each keyword value. In this case, the mapping requires specialized knowledge of the composition of the source element, and how it expands into multiple target elements.
- Many to one. The many-to-one map must specify what to do with the extra elements. If the solution adopted is to map all values of the source element to a single value in the target element, explicit rules are required to specify how concatenate the original values. Alternatively, if the solution is to map a unique value of the source element, with the consequent information loss, a rule must indicate the criteria for this value selection, e.g. the first value or the most recently added.
- Extra elements in source. Another problem arises when a source element does not have any equivalent element in the target standard. Since many metadata standards provide the ability to capture additional information or to define appropriate extensions, a rule must be established to precisely specify how these extra-elements element are handled.
- Unresolved mandatory elements in target. In some cases, mandatory elements in the target standard may have no mapping in the source standard. Because the target requires a value for the mandatory elements, the crosswalk must provide a rule to fill these elements with appropriate values.

The special rules to handle these cases would be stored in the attributes *ONE_TO_MANY, MANY_TO_ONE, EXTRA_ELEMENTS* and *UNRESOL VED_MANDATORY_ELEMENTS* of the relational table *ASSOCIATION* (see figure 3.4).

Hierarchical and structural organization

Most metadata standards organize their metadata in hierarchy of nested data structures. For instance, the FGDC CSDGM standard (FGDC, 1998) organizes the elements of the standard in sections which may be, in turn, composed of lower subsections. Working with such structured standards, the crosswalk must consider the possible differences between the hierarchies of the source and target standards. In such cases two main problems have been identified:

- Sometimes a structured section in the source standard is split up in several sections of the target standard that, although being separate, maintain some kind of relation. This relation is usually established by means of the values of some subelements of the target sections. That is to say, a foreign key constraint (similar to the foreign keys used in relational database models) must be maintained in the target standard and the value of such subelements needs to be created.
- And on the other hand, the opposite case may also occur. That is to say, elements taken from different sections in the source standard are combined to generate a target section in the target standard. This case is problematic when the source sections may have multiple instances and they are related each other by some kind of foreign keys or references.

In the process presented here, the *PATH* attribute of the relational table *ELEMENT* (shown in figure 3.4) encodes the hierarchical structure of each element in the source and target standards. This attribute contains the concatenation of the names corresponding to the broader sections (starting from the broadest section) where the element is included. And if it were necessary to specify any additional rules (e.g., for the cases specified above), they would be stored in the attribute *HIERARCHY* of the relational table *ASSOCIATION*.

3.3.4 Implementation of crosswalks: the use of style sheets

Taking into account that most metadata standards use XML as exchange and presentation format, it has been considered that the most suitable technology to carry out the implementation of crosswalks is by means of XSL (eXtensible Stylesheet Language) (W3C, 2004a), whose purpose is precisely the manipulation and transformation of XML. XSL is a language for expressing style sheets that integrates two related languages: a transformation language (XSL Transformations or XSLT) (Clark, 1999); and a formatting language (XSL Formatting Objects) of XML documents, which is comparable to the language CSS (Cascading Style Sheets) for HTML pages.

The transformation language (XSLT) provides elements that define rules to transform an XML-document into another XML-document, HTML or other text-based formats. In the case of transforming into an XML-document, this second document can use the same set of elements that the original document (it is associated to the same DTD or XML-Schema) or can use a completely different set of elements. Therefore, the method to make transformations will consist of constructing the style sheet that applied to the original XML-document (in agreement the corresponding standard of metadata) generates as a result an XML-document whose elements fulfill the target standard, and that contains the same information represented in the input document.

XSLT is a declarative match and action language. A stylesheet (XSL document) is itself an XML document that contains a set of template rules, each one consisting of a template and a pattern. Let us see a small example:

```
<xsl:stylesheet>
  <!-- ... -->
    <xsl:template match = "/rootElement/firstLevelElement/secondLevelElement" >
      <!-- actions -->
    </xsl:template>
  <!-- ... -->
</xsl:stylesheet>
```

The key element is $< xsl : template >$, which represents a template rule saying what to match and what action to take. It is applied to XML elements/attributes that match the expression contained in the *match* attribute. This match expression is defined using XPath (Clark and DeRose, 1999), a grammar for selection and navigation through the distinct parts (elements, attributes, etc.) of an XML document.

Therefore, the stylesheets implementing the crosswalks should contain templates for each one-to-one association between the sections (i.e., composite elements) of the source and the target standards, which were defined in previous phases. By one-to-one associations it is meant that an instance of the source section corresponds with a unique instance section in the target standard and can provide (contains or may access) all the values required for the elements in the target section. In such cases, the global transformation problem can be split up in small re-usable transformation sub-problems that facilitate the coding of this stylesheet. Otherwise, the transformations of elements must be applied from higher-level templates, sometimes even from the template that matches the root element of the source XML document.

If we had to construct these stylesheets by hand, the methodology for the stylesheet coding should be based in the successive creation of these templates as follows:

- Establish the document type declaration that will appear in the output document, and that will include the route (URL) of the DTD/XML-Schema corresponding to the target standard.
- Next, for each section to match in the target standard:
  - A template will be created (based on the mapping table) whose pattern is the element (name of section or subsection) in the source standard that generates the corresponding elements in the target. In this template the necessary transformation rules will be applied in order to fulfill the specification with respect to the properties and content in the target standard.
  - Once the first version of the style sheet has been built, it is applied to a XML document that conforms to the source standard, and contains values for all the elements belonging to the section previously matched. The stylesheet processor (e.g., Xalan or any other processor compliant with XSLT and XPath W3C recommendations) generates as a result a new document. Although this document will not probably validate the DTD or XML-Schema corresponding to the target standard (it only contains the sections mapped until this moment), it must be verified that the transformations have been made correctly. By means of a

XML edition tool it is possible to visualize the XML document as a tree of nodes, which correspond to the sections, subsections or simple data type elements. Therefore, this tree of nodes is used to check: the absence of a mandatory element; the order of generated elements; and the content constraints. In case of detecting some errors, the template must be revised.

– Additionally, it should be verified that there is not information loss in case the inverse style sheet were applied to the target document. Usually, a crosswalk and the inverse crosswalk are developed in parallel. If there exist some differences between the initial document and this new generated document, the mapping table should be verified to find the cause of the problem. It may be due to a problem of extra-elements in source standard that has not been resolved by any rule. But if this circumstance does not take place, the XSL template should be checked again.

– Once it has been proven that the transformation of the last section has been done correctly, the process must be started again for the next section in the source standard until the crosswalk is completely implemented.

Despite having detailed the associations between elements and the possible conversion problems, it can be supposed that the hand-coding of stylesheets is still error-prone if not done with enough thoroughness. Thus, the final aim of this implementation phase has been to automate the generation of these stylesheets as much as possible. Let us see the proposal for such automation.

Table 3.1. Mapping between elements

CSDGM				ISO19115			
Name	Path	Max	Oblig	Name	Path	Max	Oblig
Metadata	metadata	1	M	MD_Metadata	MD_Metadata	1	M
Identification Information	metadata/ idinfo	1	M	identification Info	MD_Metadata/ identi-ficationInfo	N	M
Abstract	metadata/ idinfo/ descript/ abstract	1	M	abstract	MD_Metadata/ identificationInfo/ MD_DataIdentification/ abstract	1	M
Direct Spatial Reference Method	metadata/ spdoinfo/ direct	1	M	spatial Repre-sentationType	MD_Metadata/ identificationInfo/ MD_DataIdentification/ spatialRepresentation-Type	N	OP

Comparing XML technologies and language programming compilers, one may find great similarities:

• On one hand, a compiler is defined as a program that reads a source program in one language and translates is to an equivalent language. The typical components of a compiler are a scanner (tokenizer or lexical analyzer), a parser (syntax analyzer), a semantic analyzer, an optional optimizer, a code generator, and a table of symbols. The scanner recognizes tokens, which are usually described by regular expressions. The parser verifies the syntax of the input program, being Context Free Grammars (CFG) the usual way to specify this syntax. The parser groups those tokens according to the productions of this grammar, and the application of these productions is usually represented by means of trees (parse trees or syntax trees). The semantic analyzer verifies the static semantic (data type checking, etc.) and may generate intermediate code. The optimizer improves this intermediate code that will be finally translated into target language. And additionally, it must be remarked the role of the table of symbols, which provides the mechanism to store/access the information associated with the identifiers along the compiling process.

- And on the other hand, the definition of the set of XML technologies matches up with the parts of a compiler. Both DTDs and XML-Schemas have been referred to as corresponding to different grammar models that are used to generate a set of syntax trees rather than a language. According to (Wood, 1995), a DTD can be considered as an Extended Context Free Grammar (ECFG) where the set of element types are the nonterminal and terminal symbols of the ECFG, the root element type is the initial symbol, and the element type definitions are the production rules. And other works like (Murata et al., 2001) compare DTDs and XML-Schemas with regular tree grammars, which are considered to be more appropriate for describing permissible trees than context free grammars (designed to describe permissible strings). Anyway, marked-up documents are seen as syntax trees constructed according to the grammar, where the tree structure is determined by the various tags that occur in the document and that constitute the markup. XML parsers play the role of scanners and parsers that check whether an XML document is well formed and verify the rules asserted by DTDs or XML-Schemas. Moreover, one of the possible implementations of parsers, the DOM (Document Object Model) implementation returns a tree (the syntax tree) of an input XML document. And XSLT could be compared with a code generator. The XSLT processors generate the syntax tree of the input XML document, which is later traversed any number of times applying the template rules contained in the stylesheet.

Taking into account this relation to compilers, our task is therefore the automatic construction of a code generator in the form of an XSL document. And it must be realized that this task has been implicitly performed by the work done in previous phases of the crosswalk construction process:

- The relational table *ELEMENT* (see figure 3.4) containing the description of the elements plays the role of a table of symbols. It enables the access to all the necessary information of the elements in the source standard: data types, maximum occurrence, obligation, condition and so on.

- The attribute *PATH* of the table *ELEMENT* contains the sequence of tags that is needed to access the value of an element. And it allows the construction of syntax trees representing the structure of source documents and target documents. Whereas the source syntax tree could be compared with the syntax tree usually returned by parsers, the target syntax tree dictates the structure of the output XML document.
- And finally, the *ASSOCIATION* relational table establishes the links between the nodes of the source syntax tree and the target syntax tree. That is to say, the target syntax tree establishes the tags that must be generated in the output document and the links enable the referencing to the values of the elements in the input document that must be examined.

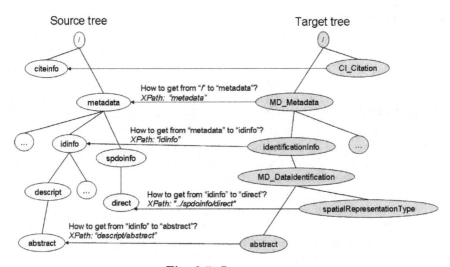

Fig. 3.5. Syntax trees

Let us see a simple example of a mapping between two standards and the corresponding syntax trees that would be generated. Table 3.1 shows the mapping that has been established between four elements of the standards CSDGM and ISO 19115 [15]. And figure 3.5 displays the corresponding syntax trees of both standards and the cross links between the nodes of these trees. Additionally, the figure shows the XPath expressions that the envisioned XSL document must include to traverse the input XML document and access the next value of a source element. For instance, after writing the *identification-Info* tag in the output document the XSL is positioned at an *idinfo* element

---

[15] This example is only illustrative. It does not pretend to be exhaustive at all, and other standards could have been used instead. The mapping between CSDGM and ISO 19115 is described later in section 3.4.1.

in the input document. Then, the stylesheet must write the subelements of *identificationInfo* element. Thus, according to the structure of the target tree, it writes the *MD_DataIdentificationInfo* tag and its child tag *abstract*. At this moment, it needs the XPath expression to traverse the input XML document until reaching the source *abstract* element. This XPath expression is the path between the source nodes that are linked with the target node (whose value is needed) and the closer parent node of this target node being linked with the source tree. In this case, it is necessary to obtain the path between the node labeled as *idinfo* (linked with the first parent in the target tree having a link with the source tree) and the one labeled as *abstract* (linked with the target node). This path concatenates the labels of the nodes in case of moving forwards (e.g., *"descript/abstract"* for moving from *idinfo* node to *abstract* node) and *".."* in case of moving backwards. For instance, in order to write the value of the *spatialRepresentationType* element, the XPath expression *"../spdoinfo/direct"* must be applied to move from the *idinfo* element in the input XML document.

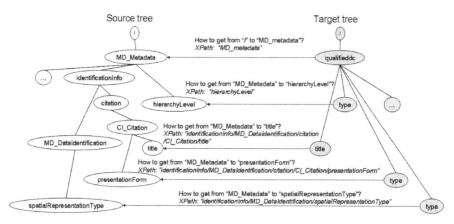

Fig. 3.6. Different source elements to a target element

It is also worth mentioning a frequent case in these cross linked syntax trees that arises when a target element is mapped to different source elements or viceversa. The options here would be to establish multiple links from the target element to the source elements or to replicate the target element and establish a separate link with each source element. Given that special conversions could be needed for each mapping or that the source and the target elements may be composed of nested elements, it has been finally decided to replicate the target elements and have separate mappings. This will clarify the construction of the desired XSL document. For instance, figure 3.6 shows the mapping between two elements (*title* and *type*) of Dublin Core and the corresponding elements of ISO 19115. The Dublin Core element *type* is mapped with three elements of ISO 19115: *MD_Metadata/hierarchyLevel*, *MD_Metadata/identificationInfo/MD_DataIdentification/ spatialRepresentation-Type*, and *MD_Metadata/identificationInfo/citation/CI_Citation/ presentationForm*.

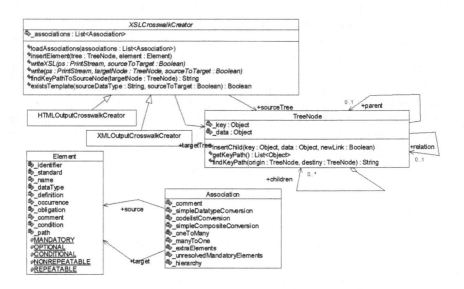

Fig. 3.7. Stylesheet writing

Thus, given the availability of these syntax trees and their cross-links, it is possible to write an algorithm that will traverse in parallel both trees and will create an initial version of the stylesheet. Figure 3.7 shows the classes that take part in this algorithm. The *TreeNode* class represents a tree whose nodes store the elements (the *data* attribute) of a metadata standard and are labeled (the *key* attribute) with its XML tag. These nodes may have an unlimited number of child nodes and they can also have a *relation* with other *TreeNode* instance. This last relation enables the links between nodes of source and target trees. Classes *Element* and *Association* are in-memory representations of the descriptions of elements and associations stored as tuples of the relational tables *ELEMENT* and *ASSOCIATION* (see figure 3.4). And finally, the *XSLCrosswalkCreator* class is in charge of writing the XSL stylesheet. The *loadAssociations* method of this class loads a list of associations between elements and constructs the cross-linked source and target syntax trees. And the method *writeXSL* is in charge of writing the two possible stylesheets: the source-to-target stylesheet (boolean parameter *sourceToTarget* is equals to *true*) or the reverse stylesheet from target to source (boolean parameter *sourceToTarget* is equals to *false*). It must be noted that this last method is abstract as well as the *XSLCrosswalkCreator* class. This is motivated by the fact that the *XSLCrosswalkCreator* class can be specified in different classes according to the output format. *XMLOutputCrosswalkCreator* is the derived class that will generate XML output. But one could also be interested in other types of output. For instance, *HTMLOutputCrosswalkCreator* would transform the input XML into a target standard/schema, which is only displayed in HTML format.

```
1:   public void writeXSL(PrintStream ps, boolean sourceToTarget){
2:      ps.println(VERSION);
3:      ps.println(BEGIN_STYLESHEET);
4:      ps.println(OUTPUT);
5:      TreeNode targetTree = (sourceToTarget)?getTargetTree():getSourceTree();
6:      ListIterator it = targetTree.getChildren().listIterator();
7:      // write templates for root and reusable elements
8:      while (it.hasNext()) {
9:        TreeNode child =(TreeNode)it.next();
10:       ps.println(beginTag(TEMPLATE,MATCH,findKeyPathToSourceNode(child)));
11:       write(ps,child,sourceToTarget);
12:       ps.println(endTag(TEMPLATE));
13:      }
14:      ps.println(END_STYLESHEET);
15:  }
```

Fig. 3.8. The *writeXSL* method of *XMLOutputCrosswalkCreator*

Figure 3.8 shows the *writeXSL* method, which is in charge of writing the XSL document. Apart from the aforementioned *sourceToTarget* boolean parameter, it has another parameter (*ps* of type *PrintStream*) that references the output stream where the stylesheet is written. As it can be observed, this method writes the initial instructions of the stylesheet document and creates the templates that will generate the target elements corresponding with the nodes directly linked with the root node (labeled as /) in the target syntax tree. For the sake of simplicity, methods like *beginTag*, *endTag* or *completeTag* and constants (public static final attributes in Java) like *VERSION*, *OUTPUT*, *TEMPLATE* or ELEMENT facilitate the writing of XML tags and the final strings appearing in the output document (e.g., $< xsl : output\ method = "xml"\ indent = "yes"\ encoding = "ISO - 8859 - 1" >$ for the case or *OUTPUT*). In order to write the content of these templates, this method makes use of the *write* method at line 11.

Figure 3.9 displays the *write* method that has the responsibility of writing a target element. It is a recursive method that makes recursive invocations in case of dealing with complex elements (intermediate nodes in the tree) that are composed of subelements. Lines 3-12 deal with the base case, i.e. the case of a leaf in the tree. In this situation, two possibilities may arise to obtain the value of the element: the value of the element is directly obtained from the source element (lines 10-12); or there is a special template that must be applied to obtain the value of this element (lines 8-9) from the input document (e.g., mapping between *citeinfo* and *CI_Citation* in figure 3.5). The

```
 1: public void write(PrintStream ps, TreeNode targetNode, boolean sourceToTarget){
 2:   ps.println(beginTag(ELEMENT,NAME,targetNode.getKey()));
 3:   if (targetNode.getChildren().isEmpty())
 4:   { // base case:we have found a leaf of a tree and we must write its value
 5:     Element sourceElement = (Element)targetNode.getRelation().getData();
 6:     if (sourceElement!=null)
 7:       if (existsTemplate(sourceElement.getDataType(),sourceToTarget))
 8:         //The dataType of this element is complex and there is a separate template.
 9:         ps.println(completeTag(APPLYTEMPLATES,SELECT,sourceElement.getDataType()));
10:       else
11:         // write source node value
12:         ps.println(completeTag(VALUEOF,SELECT,"."));
13:   } else { // recusive step: children must be recursively written
14:     ListIterator it = targetNode.getChildren().listIterator();
15:     while (it.hasNext())  {
16:       TreeNode child = (TreeNode) it.next();
17:       if (child.getRelation()!=null) {
18:         // obtain keyPath from source node connected with 'targetNode'
19:         // until source node connected with 'child'
20:         String keyPath = findKeyPathToSourceNode(child);
21:         Element targetElement = (Element)child.getData();
22:         // check mandatory constraints
23:         boolean checkMandatory = (targetElement!=null&&
24:                        targetElement.getObligation()==Element.MANDATORY);
25:         if (checkMandatory) {
26:           // write begin of choose clause
27:           ps.println(beginTag(CHOOSE));
28:           ps.println(beginTag(WHEN,TEST,keyPath));
29:         }
30:         // assure constraint of maximum number of occurrences
31:         if (targetElement!=null&&targetElement.getOccurrence()==
32:         Element.NONREPEATABLE) // only first instance is allowed
33:           ps.println(beginTag(FOREACH,SELECT, keyPath+"[1]"));
34:         else // manage all instances
35:           ps.println(beginTag(FOREACH,SELECT, keyPath));
36:         write(ps,child,sourceToTarget);
37:         ps.println(endTag(FOREACH));
38:         if (checkMandatory) {
39:           // write end of choose clause
40:           ps.println(endTag(WHEN));
41:           ps.println(beginTag(OTHERWISE));
42:           ps.println(beginTag(ELEMENT,NAME,child.getKey()));
43:           ps.println("mandatory element without value");
44:           ps.println(endTag(ELEMENT));
45:           ps.println(endTag(OTHERWISE));
46:           ps.println(endTag(CHOOSE));
47:         }
48:       } else
49:         // There is not a linked element in the source. It is not necessary
50:         // to move within the input XML document (write for-each sentences)
51:         write(ps,child,sourceToTarget);
52:     }
53:   }
54:   ps.println(endTag(ELEMENT));
55: }
```

Fig. 3.9. The *write* method of *XMLOutputCrosswalkCreator*

method *existsTemplate* of *XSLCrosswalkCreator* is the one in charge of verifying whether there is a template to write the content of an element or not.
A template must be applied whenever there is a node at the first level of the
source tree labeled with the data type of the source element[16]. And on the
other hand, lines 13-55 deal with the case of intermediate nodes in the tree,
i.e. nodes corresponding to complex elements that are composed of subelements. In this case, we will need to treat each subelement and make recursive
invocations to the *write* method. In addition, it must be checked whether the
subelements have a link in the source tree or not (line 17). If there is a link,
before making the recursive invocation it is necessary to add a *for instruction*
($< xsl : for - each\ select = "path" >$) that will traverse the input XML document to reach the appropriate element (see lines 30-37). This path is given
by the *findKeyPathToSourceNode* method in *XSLCrosswalkCreator* class. Finally, it must be observed that the algorithm also takes into account the
constraints related with the repeatability and the obligation of an element.
The obligation of an element is managed by means of a *switch instruction*
($< xsl : choose... >$) in lines 25-29 and 38-47. In case of not finding a value
in the input XML document, the subelement is filled with a warning message
(see lines 38-47) in the *otherwise* branch ($< xsl : otherwise >$) of the *switch*
instruction. Last, the repeatability of an element is controlled in line 31. If
the target element is non-repeatable, the *for* instruction traversing the input
XML document will take only the first occurrence of the related element. Otherwise, all the occurrences of the source element are translated to the target
element.

Finally, figure 3.10 shows the source-to-target stylesheet that is automatically generated for the example in figure 3.5. It must be remarked that this
stylesheet is only an initial version. At the moment, additional rules are not
considered by the *XMLOutputCrosswalkCreator*. These rules depend on each
particular case and they are difficult to automate. But at least, this automatic
generation of stylesheets reduces the hand-coding to just a few places.

3.4 Putting the method to work

Following the process explained in previous section, several crosswalks have
been developed in order to validate it and to make possible the interoperability among several metadata standards: the CSDGM of FGDC (FGDC,
1998), ISO 19115 (ISO, 2003a) and Dublin Core (ANSI, 2001; DCMI, 2004;
ISO, 2003d). As it can be observed two of them, CSDGM and ISO 19115,
are genuine geographic metadata standards. But it has been also tested the
interoperability between these standards and general-purpose standard like
Dublin Core, which is used across very different domains.

---

[16] The nodes at the first level of the syntax tree correspond to the initial element
of a metadata record or to a reusable section (a complex data type).

```
<?xml version="1.0" encoding="ISO-8859-1"?> <xsl:stylesheet
version="1.0" xmlns:xsl="http://www.w3.org/1999/XSL/Transform">
<xsl:output method="xml" indent="yes" encoding="ISO-8859-1"/>
<xsl:template match="metadata" >
 <xsl:element name="MD_Metadata" >
  <xsl:for-each select="idinfo" >
   <xsl:element name="identificationInfo" >
    <xsl:element name="MD_DataIdentification" >
     <xsl:choose>
      <xsl:when test="descript/abstract" >
       <xsl:for-each select="descript/abstract[1]" >
        <xsl:element name="abstract" ><xsl:value-of select="." /></xsl:element>
       </xsl:for-each>
      </xsl:when>
      <xsl:otherwise>
       <xsl:element name="abstract" >mandatory element without value</xsl:element>
      </xsl:otherwise>
     </xsl:choose>
     <xsl:for-each select="../spdoinfo/direct" >
      <xsl:element name="spatialRepresentationType" ><xsl:value-of select="." />
      </xsl:element>
     </xsl:for-each>
    </xsl:element>
   </xsl:element>
  </xsl:for-each>
 </xsl:element>
</xsl:template>
</xsl:stylesheet>
```

Fig. 3.10. Example of generated stylesheet

In order to illustrate the construction of these crosswalks, this section will provide an overview of two of these transformations: CSDGM ↔ ISO 19115, and Dublin Core ↔ ISO 19115.

3.4.1 Transformation between CSDGM and ISO 19115

Although the purpose of these two standards is to describe a geographic information resource, they present some important differences. With respect to the documentation and organization of the standard, CSDGM standard (see left side of figure 3.11) is structured in 10 sections (7 main sections and 3 reusable sections) and contain 469 different elements, from which 119 are composite elements (their existence is justified to contain other elements). The syntax of the standard is expressed by means of BNF production rules and in addition to this, a definition for the content of each element is also provided. On the opposite, ISO standard (see right side of figure 3.11) uses object-oriented methodology and it is specified by means of the use of UML diagrams, which model the relations and organization of the information captured in this standard. Besides, ISO standard provides a data dictionary that gathers the names, descriptions, and domain constraints of all classes and attributes, 509 elements altogether. Given this situation, CSDGM main sections could be compared with the ISO packages, which compile the different classes representing the meta-information captured by ISO standard.

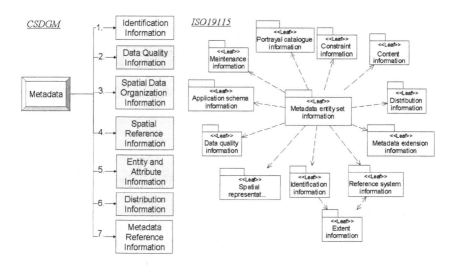

Fig. 3.11. Metadata models of CSDGM(left) and ISO 19115(right)

Table 3.2. Mapping between CSDGM and ISO 19115 sections

CSDGM Section	ISO Package
Main Sections	
Identification information	Identification information (including the references to the sections Constraint information, Maintenance information)
Data quality information	Data quality information
Spatial data organization Information	Spatial representation information
Spatial reference information	Reference System Information
Entity and Attribute Information	Content information
Distribution Information	Distribution Information
Metadata Reference Information	Metadata entity set information (including references to the Constraint information, Maintenance information, and Metadata extension information sections)
	Portrayal catalogue information
	Application schema information
Reusable sections	
Citation information, Contact Information	Citation and responsible party information
Time period information	

Table 3.2 displays the mapping between sections and packages at a higher level. Although this direct mapping does not necessarily exist for deeper levels, analogous elements can be found at different points of the hierarchy.

Regarding semantic information, the ISO standard, thanks to its recent appearance and its conciliating character, resolves some deficiencies that can be found in CSDGM standard. For example, ISO standard provides the data types *raster* and *imagery*, whereas in CSDGM there is only the first one. Moreover, these standards present slight differences in the terminology. For instance, the element bounding box of the CSDGM standard contains four coordinate elements, whose short names are *westbc*, *eastbc*, *northbc* and *southbc*.

| | | CSDGM (FGDC) | | | | | | | | | ISO19115 | | | | |
| | | Name | | | Obli | Max | Path | Identifier | | | Name | | | Obli | Maxi |
Ident	1	2	3	4			(last tag)		1	2	3	4	5		
8.0	Citation_Information						citeinfo	361.	CI_Citation						
								369.:376.		CitedResponsibleParty:CI_ResponsibleParty				O	N
8.1		Originator			M	N	origin	378.			OrganisationName			C	1
								380.:389.			ContactInfo:CI_Contact			O	1
								392.:4				OnLineResource:CI_OnLineResource		O	1
8.10		Online_Linkage			O	N	onlink	5.					Linkage	M	1
								381.				Role		M	1
													CI_RoleCode		
								364.:1.			Date:CI_Date			M	N
8.2		Publication_Date			M	1	pubdate	2.				Date		M	1
		Publication_Time			O	1	pubtime								
								3.				DateType		M	1
													CI_DateTypeCode		
8.4		Title			M	1	title	362.			Title			M	1
								363.			AlternateTitle			O	N
8.5		Edition			C	1	edition	365.			Edition			O	1
								366.			EditionDate			O	1
8.6		Geospatial_Data_Presentatio			C	1	geoform	370.			PresentationForm			O	N
												CI_PresentationFormCode			
8.7		Series_Information			C	1	serinfo	371.:11			Series: CI_Series			O	1
8.7.1			Series_Name		M	1	sername	12.				Name		O	1
8.7.2			Issue_Identification		M	1	issue	13.				IssueIdentification		O	1
								14.				Page		O	1
8.8		Publication_Information			C	1	pubinfo	369.:376.			CitedResponsibleParty:CI_ResponsibleParty			O	N
								380.:389.			ContactInfo:CI_Contact			O	1
								391.:382.				Address:CI_Address		O	1
8.8.1			Publication_Place		M	1	pubplace	383.					City	O	1
8.8.2			Publisher		M	1	publish	378.				OrganisationName		C	1
								381.				Role		M	1
													CI_RoleCode		
8.9		Other_Citation_Details			C	1	othercit	372.			OtherCitationDetails			O	1
8.11		Larger_Work_Citation			C	1	lworkcit								
			Citation_Information		M	1	citeinfo								
								367.			Identifier			O	N
								368.			IdentifierType			O	N
								373.			CollectiveTitle			O	1
								374.			ISBN			O	1
								375.			ISBN			O	1

Table 3.3. Detailed mapping between CSDGM and ISO 19115

The corresponding element in the ISO standard also contains four elements but this time the short names are *westBL*, *eastBL*, *northBL* and *southBL*. The only difference between these elements consists in a question of terminology, as they are semantically equivalent.

Despite the differences, one of the commonalities in both standards is the fact that the most accepted format for exchange and encoding is XML. The only way to assure that an XML-document is compliant with the standard is validating this document against the DTD provided by the organization that defined the standard. Therefore, a crosswalk implementation based on style sheets is the most accurate solution. The CSDGM → ISO style sheet that has been created enables the transformation of five of the seven main sections of the CSDGM. That includes all the mapping of sections that are mandatory in both standards. The two sections that have not been matched yet are *Spatial_Reference_Information* and *Entity_and_Attribute_Information*. The matching was not possible because their organization and conception were absolutely disparate in both standards. For instance, whereas the names of the subsections of *Spatial_Reference_Information* of the CSDGM correspond with the different coordinate systems (Transverse Mercator, Mercator, Equidistant Conic,...), ISO uses the codes maintained by recognized organizations

```
<?xml version = '1.0' encoding = 'ISO-8859-1'?> <!DOCTYPE metadata
SYSTEM "http://www.fgdc.gov/metadata/fgdc-std-001-1998.dtd">
<metadata>
  <idinfo>
    ...
    <citation>
     <citeinfo>
       <origin>National Imagery and Mapping Agency</origin>
       <pubdate>2000-09-03</pubdate>
       <title>VMAPLV0</title>
       <geoform>mapDigital</geoform>
       <pubinfo>
         <pubplace>Bethesda, United States</pubplace>
         <publish>National Imagery and Mapping Agency</publish>
       </pubinfo>
       <othercit>Vector Map: a general purpose database design to support GIS
          applications</othercit>
     </citeinfo>
    </citation>
    ...
  </idinfo>
  ...
</metadata>
```

Fig. 3.12. Original CSDGM metadata in XML

(e.g. European Petroleum Survey Group[17]), which maintain an updated catalog of coordinate systems, ellipsoids or datums and whose citation is also included within metadata. Concerning the *Entity_and_Attribute_Information*, the problem is that ISO only stores a brief description of features, the equivalent concept to entities in CSDGM. All the information about the attributes of entities and their values have no place in ISO 19115 metadata model.

Finally, the transformation of the subsection *Citation_Information* of CSDGM will be shown as an example of the crosswalk. Table 3.3 shows the mapping table between the CSDGM *Citation_Information* section and the ISO *CI_Citation* entity.

Appendix B.1 shows a piece of the CSDGM → ISO 19115 XSL stylesheet that implements the crosswalk according to the previous table. Figure 3.12 and 3.13 show the effect of applying the previous stylesheet. Figure 3.12 contains an extract of an XML metadata file in conformance with CSDGM and figure 3.13 displays the piece of XML conforming to ISO 19115, which has been obtained after applying the stylesheet.

One of the main conclusions about the mapping is that these standards are very similar and thus, most metadata have been successfully translated. This is not surprising because ISO 19115 is a consensus standard that has been taken into account previous standards and CSDGM was perhaps the most important. The FGDC standard was created in 1998 and since then it has been used by many professionals in the GIS domain. Furthermore, FGDC members have participated actively in the construction of ISO 19115.

---

[17] http://www.epsg.org/

```
<?xml version = '1.0' encoding = 'ISO-8859-1'?>
<iso19115:MD_Metadata xmlns:iso19115="http://www.isotc211.org/iso19115/">
    <identificationInfo>
        <iso19115:_MD_Identification xmlns:xsi="http://www.w3.org/2001/XMLSchema-instance"
            xsi:type="iso19115:MD_DataIdentification">
        <citation>
            <title>VMAPLV0</title>
            <date>
                <date>2000-09-03</date>
                <dateType>publication</dateType>
            </date>
            <citedResponsibleParty>
                <organisationName>National Imagery and Mapping Agency</organisationName>
                <role>originator</role>
            </citedResponsibleParty>
            <citedResponsibleParty>
                <organisationName>National Imagery and Mapping Agency</organisationName>
                <contactInfo>
                    <address>
                        <city>Bethesda, United States</city>
                    </address>
                </contactInfo>
                <role>publisher</role>
            </citedResponsibleParty>
            <presentationForm>mapDigital</presentationForm>
            <otherCitationDetails>Vector Map: a general purpose database design to
                support GIS applications</otherCitationDetails>
        </citation>
                    ...
        </iso19115:_MD_Identification>
    </identificationInfo>
    ...
</iso19115:MD_Metadata>
```

Fig. 3.13. Derived ISO 19115 metadata in XML

3.4.2 Transformation between ISO 19115 and Dublin Core

As it was mentioned in section 1.4.2, the European Committee for Standardization (CEN) has developed a project, implemented as a workshop, to provide an open forum in which Dublin Core metadata standards related issues get addressed, specifically in support of present and future projects (e.g., those concerned with Information Society Technology programs). This workshop is the CEN/ISSS Workshop "Metadata for Multimedia Information - Dublin Core (MMI-DC)" and has developed a work package, the work Item 7 - "Define and agree a CEN Workshop Agreement (CWA) on mappings between Dublin Core and the forthcoming ISO 19115 standard for geographic information metadata", whose three main deliverables are: a crosswalk between the standard ISO 19115 and Dublin Core, guidance material for the use of this crosswalk, and a spatial application profile to extend Dublin Core for describing geographic information resources. The authors of this book have participated in this work item (Zarazaga-Soria et al., 2003d,c,b), which was finished in September 2003. In particular, this section presents a subset of this work focused on the mapping between ISO 19115 and Dublin Core, which was done following the methodology proposed in this chapter.

Table 3.4. Dublin Core - ISO 19115 Core mapping

DC element	ISO-CORE element
TITLE	Dataset title (M) (MD_Metadata.identificationInfo.citation.title)
CREATOR	Dataset responsible party (O) (MD_Metadata.identificationInfo. pointOfContact, role="originator")
SUBJECT	Dataset topic category (M) (MD_Metadata.identificationInfo. topic-Category)
DESCRIPTION	Abstract describing the dataset (M) (MD_Metadata.identificationInfo. abstract)
PUBLISHER	Dataset responsible party (O) (MD_Metadata.identificationInfo. pointOfContact, role="publisher") Metadata point of contact (M) (MD_Metadata.contact)
DATE	Dataset reference date (M) (MD_Metadata.identificationInfo.citation. date) Metadata date stamp (M) (MD_Metadata.dateStamp)
TYPE	Spatial representation type (O) (MD_Metadata.identificationInfo. spatialRepresentationType)
FORMAT	Distribution format (O) (MD_Metadata.distributionInfo. distribution-Format)
IDENTIFIER	On-line resource (O) (MD_Metadata.distributionInfo. transferOptions.onLine.linkage)
SOURCE	Lineage (O) (MD_Metadata.dataQualityInfo.lineage. source.description)
LANGUAGE	Dataset language (M) (MD_Metadata.identificationInfo.language)
COVERAGE (refinement *spatial*) (refinement *temporal*)	 Geographic location of the dataset (by four coordinates or by geographic identifier) (C) (MD_Metadata.identificationInfo. extent.geographicElement) Additional extent information for the dataset (vertical and temporal) (O) (MD_Metadata.identificationInfo. extent.temporalElement.extent)

As mentioned in previous sections, ISO 19115 is mainly oriented to the description of digital data (geographic datasets, dataset series or individual geographic features) but its principles may be also extended for many other forms of geographic data such as maps, charts, and textual documents as well as non-geographic data. Another remarkable aspect concerning ISO 19115 is that despite defining an extensive set of metadata elements, in practice only a subset of these elements is used. However, it is essential to maintain a basic minimum number of metadata elements for describing geographic datasets. For this purpose, the standard has defined a profile entitled as "Core metadata for geographic datasets" that comprises a small list of core metadata elements (22 elements). These core metadata elements facilitate interoperability because they allow users to understand without ambiguity the geographic data and metadata provided by either producers or distributors. Furthermore, ISO 19115 enforces all application profiles of this standard to include these core elements.

On the other hand, Dublin Core does not aim at displacing other metadata standards. Instead, it is intended to co-exist (frequently Dublin Core descriptors form part of broader resource descriptions) with metadata standards that offer other semantics. In fact, the potential of Dublin Core is to provide the visibility of a collection of resources across different subject domains and at a low cost.

Therefore, for standards like ISO 19115 which do not describe geographic information as general-purpose data, the interoperability with Dublin Core results very appealing. The tool to facilitate this interoperability is the definition of automatic-crosswalks between these standards. Following the process described above, the crosswalk between both standards has been built. And as the ISO 19115 "Core metadata for geographic datasets" compiles the 22 elements that minimally describe a geographic resource, this crosswalk is mainly focused in the mapping between Dublin Core and these basic elements of ISO. Table 3.4 presents the mapping table between Dublin Core and ISO 19115 Core.

```
<?xml version="1.0" encoding="ISO-8859-1"?>
<iso19115:MD_Metadata xmlns:iso19115="http://www.isotc211.org/iso19115/" ... >
    <iso19115:_MD_Identification xsi:type="iso19115:MD_DataIdentificationType">
        <citation>
            <title>VMAPLV0</title>
            ...
        </citation>
        ...
        <pointOfContact>
            <contactInfo>
                <address>...</address>
                <onlineResource>...</onlineResource>
            </contactInfo>
            <role><CI_RoleCode_CodeList>originator</CI_RoleCode_CodeList></role>
            <organisationName>National Imagery and Mapping Agency</organisationName>
            <positionName> Director, NIMA, ATTN:COD, MS P-37</positionName>
        </pointOfContact>
        ...
    </iso19115:_MD_Identification>
    ...
</iso19115:MD_Metadata>
```

Fig. 3.14. ISO 19115 metadata file example

```
<?xml version = '1.0' encoding = 'ISO-8859-1'?>
<rdf:RDF xmlns:rdf="http://www.w3.org/1999/02/22-rdf-syntax-ns#">
    <rdf:Description>
        <dc:title xmlns:dc="http://purl.org/dc/elements/1.1">VMAPLV0</dc:title>
        <dc:creator xmlns:dc="http://purl.org/dc/elements/1.1">
        National Imagery and Mapping Agency</dc:creator>
        ...
    </rdf:Description>
</rdf:RDF>
```

Fig. 3.15. Generated Dublin Core metadata file

Appendixes B.2 and B.3 show an extract of the XSL stylesheets implementing the crosswalks ISO 19115→DC and DC→ISO 19115 respectively, which are derived from the mapping tables presented in this section. For the sake of clarity, only the transformation of the elements "DC:TITLE" (the ISO "MD_Metadata. identificationInfo.citation.title") and "DC:CREATOR" (the

Table 3.5. Dublin Core elements that must be mapped to ISO 19115 Comprehensive (no match with ISO 19115 core)

DC element	ISO 19115 Comprehensive
CONTRIBUTOR	MD_Metadata.identificationInfo.credit
RELATION (no refinement or using *isVersionOf, replaces, isPartOf, references, isFormatOf*) (refinement *isPartOf*)	MD_Metadata.identificationInfo. aggregationInfo MD_Metadata.identificationInfo. citation.series.name
RIGHTS (no refinement) (refinement *accessRights*)	 MD_Metadata.identificationInfo. resourceConstraints MD_Metadata.identificationInfo.resourceConstraints. accessConstraints
AUDIENCE (no refinement or using *educationLevel*) (refinement *mediator*)	MD_Metadata. identificationInfo.purpose MD_Metadata.distributionInformation.distributor. distributorContact (role="distributor")

Table 3.6. Description of the ISO 19115 Comprehensive elements in table 3.5

ISO 19115 element	Description
MD_Metadata.identificationInfo. credit	Recognition of those who contributed to the resource(s)
MD_Metadata.identificationInfo. aggregationInfo	Aggregate dataset information
MD_Metadata.identificationInfo. citation.series.name	name of the series, or aggregate dataset, of which the dataset is a part
MD_Metadata.identificationInfo. resourceConstraints	Constraints describe constraints applied to assure the protection of privacy or intellectual property, and any special restrictions or limitations or warnings on using the resource
MD_Metadata.identificationInfo. resourceConstraints.accessConstraints	Constraints applied to assure the protection of privacy or intellectual property, and any special restrictions or limitations on obtaining the resource
MD_Metadata.identificationInfo. purpose	Summary of the intentions with which the resource(s) was developed
MD_Metadata. distributionInformation. distributor.distributorContact (when role=distributor)	Party from whom the resource may be obtained. This list need not be exhaustive

ISO "MD_Metadata. identificationInfo.pointOfContact") has been displayed. The complete version of the stylesheets can be found in (Zarazaga-Soria et al., 2003c). ISO 19115→ DC stylesheet takes as input an XML file in conformance with the XML-Schema provided in (ISO, 2003b) and produces a Dublin Core metadata file encoded in RDF/XML (Dublin Core metadata is frequently exchanged in RDF, see section 1.4.2). And on the contrary, DC→ ISO 19115 stylesheet receives an RDF/XML file with one or more RDF descriptions ("rdf:Description" tags). However, as the output is an ISO 19115 metadata file and this file must contain the description of a unique geographic information resource, only the fist occurrence of "rdf:Description" will be taken into account. Figure 3.14 contains a piece of an ISO 19115 metadata file and figure 3.15 shows the generated Dublin Core metadata file after applying the ISO 19115→DC stylesheet.

Table 3.7. Possible solution for ISO 19115 Core elements with no Dublin Core mapping

ISO-CORE element	Description and mapping
Dataset character set (C) (MD_Metadata.identificationInfo. characterSet)	This is the full name of the character coding standard used for the dataset. The mapping between Dublin Core and ISO 19115 Core can be implemented as a refinement of the "FORMAT" Dublin Core element.
Spatial resolution of the dataset (O) (MD_Metadata.identificationInfo. spatialResolution)	This is the factor which provides a general understanding of the density of spatial data in the dataset. The mapping between Dublin Core and ISO 19115 Core cannot be implemented because this is a specific geographic feature of the resource. One possible solution is the extension of the Dublin Core element set.
Reference system (O) (MD_Metadata. referenceSystemInfo)	This term provides information about the reference system. The mapping between Dublin Core and ISO 19115 Core cannot be implemented directly because this is a specific geographic feature of the resource. One possible solution is the extension of the Dublin Core element set.
Metadata file identifier (O) (MD_Metadata.fileIdentifier)	This element represents the unique identifier for this metadata file. A possible mapping of this ISO 19115 Core element could be its definition as a refinement of the "IDENTIFIER" Dublin Core element. However, it may result complex to generate a unique identifier for metadata descriptions, particularly if data and metadata are delivered separately.
Metadata standard name (O) (MD_Metadata. metadataStandardName)	This term stores the name of the metadata standard (including profile name). It has no mapping with any Dublin Core element. However, the standard name could be auto generated for a mapping from DC to ISO. The objective is precisely to obtain metadata compliant with ISO 19115 Core, i.e. metadata using ISO 19115 as standard name. On the other hand, regarding Dublin Core metadata descriptions, the encoding itself should reference the document defining the Dublin Core elements.
Metadata standard version (O) (MD_Metadata. metadataStandardVersion)	This term stores the version (profile) of the metadata standard used. It has no mapping with any Dublin Core element. However, the standard name could be auto generated for a mapping from DC to ISO. The objective is precisely to obtain metadata compliant with ISO 19115 Core and a specific version. On the other hand, regarding Dublin Core metadata descriptions, the encoding itself should reference the document and version that defines the Dublin Core elements.
Metadata language (C) (MD_Metadata.language)	This term keeps the language used for documenting metadata. The mapping between Dublin Core and ISO 19115 Core can be implemented as a refinement of the "LANGUAGE" Dublin Core element.
Metadata character set (C) (MD_Metadata.characterSet)	This element represents the full name of the character coding standard used for the metadata set The mapping between Dublin Core and ISO 19115 Core can be implemented as a refinement of the "FORMAT" Dublin Core element.

In order to establish the syntax of RDF/XML documents containing Dublin Core metadata, there are two possible mechanisms: a DTD (Document Type Definition) or an XML-Schema (an enhanced version of a DTD). Regarding Simple Dublin Core metadata, the document "Expressing Simple Dublin Core in RDF/XML" [18] includes a DTD [19] and an XML-Schema[20] that

---

[18] A DCMI recommendation available at http://dublincore.org/documents/2002/ 07/31/dcmes-xml/.

[19] http://dublincore.org/documents/2002/07/31/dcmes-xml/dcmes-xml-dtd.dtd

[20] http://dublincore.org/schemas/xmls/simpledc20021212.xsd

define the syntax for expressing simple Dublin Core metadata, i.e. without qualifiers, in RDF/XML. And as concerns Qualified Dublin Core, DCMI published a document entitled "Expressing Qualified Dublin Core in RDF/XML" [21]. Nevertheless, this document does not include any kind of DTD or XML Schema. XML schemas for Qualified Dublin Core are currently under development.

Finally, as a result of the crosswalk construction, it has been observed that there are four elements of Dublin Core that have no correspondence with any element of the Core version of the ISO 19115. These four elements are *CONTRIBUTOR, RELATION, RIGHTS* and *AUDIENCE*. Nevertheless, all of them have a correspondence with one or more elements of the ISO 19115 Comprehensive profile. The ISO 19115 Comprehensive profile fully defines the complete range of metadata required to identify, evaluate, extract, employ, and manage geographic information. In fact, it almost includes all the metadata entities defined in the ISO 19115 document. Tables 3.5 and 3.6 show this mapping to the ISO elements contained in the Comprehensive profile. The lack of mapping between these last 4 DC elements and ISO 19115 Core metadata could justify the expansion of the ISO 19115 Core Metadata to include the comprehensive elements appearing in table 3.5. This way, a full mapping DC → ISO 19115 Core would be possible. The aim of Dublin Core is to compile the minimum elements that describe a resource and thus ISO Core should include at least these Dublin Core elements to be really "Core".

Additionally, another deficiency in the mapping that can be observed is that there are some elements from the Core version of the ISO 19115 having no direct correspondence with elements from Dublin Core. Table 3.7 presents these elements and proposes a solution for their mapping (if it exists). Furthermore, the mapping described in this section and the deficiencies observed have motivated the creation of a spatial application profile of Dublin Core (Zarazaga-Soria et al., 2003b).

3.5 Conclusions and future work

This chapter has presented the process followed to carry out the construction of a series of crosswalks that enable the interoperation between different metadata schemas. This process consists of a series of steps that gradually incorporate fine grained details about the source-to-target mapping until the full crosswalk is finished. Thanks to this process, it is possible to establish a semi-formalized method that implies a rigorous specification of standards and transformation, thus minimizing the possible loss of information. Furthermore, it must be remarked that this process is context free and thus, it can be applied to transform metadata in any domain context or even to transform source and metadata schemas from different domains.

---

[21] Proposed recommendation available at http://dublincore.org/documents/2002/04/14/dcq-rdf-xml/.

There are two main reasons that have motivated the necessity of crosswalks: the convergence towards international standards and the reusability of resources across different domains. Along the last decade and as a response to the uncontrolled diffusion of multimedia objects encoded in disparate formats, many organizations (standardization bodies, software vendors, ...) started different initiatives for the definition of metadata standards to enable the common understanding within a community of users. However, despite the initial intention of common understanding, the diversity of initiatives originated also an undesired effect of heterogeneity. Now, most of these initiatives have converged to international standard but the legacy metadata (the work done in the past) can not be directly thrown away. This is clearly the case of geographic metadata standards where different standards like CSDGM (FGDC, 1998) or CEN/TC 287 prENV 12657 (CEN, 1998) aim at migrating towards the international ISO 19115 (ISO, 2003a) standard. The second reason arises from the necessity of facilitating search of resources across different domains. Although digital libraries may be specialized on particular types of resource and use specific metadata for such resources, they are also asked to provide general descriptions of their resources for the sake of interoperability. For instance, spatial data infrastructures and geolibraries, apart from using ISO 19115 metadata, should provide a summary view of their specific geographic metadata, understandable by general public or discovery agents. This summary view could be the one defined by Dublin Core, which has great acceptance in public administration or in the description of web resources. Under these requirements, the creation of multiple versions of the same metadata to facilitate this multiple-standard visibility does not prove to be the best option. On the contrary, a more sensible option would be to maintain the metadata in the original standard and apply the necessary crosswalks when other views are required.

Additionally, this chapter has presented the applicability of this crosswalk methodology to enable the interoperability among several metadata standards, mainly in the geographic information domain but also providing interoperation with the general purpose Dublin Core standard. As an example of the interoperability in the geographic information domain, some details about the crosswalk between CSDGM and ISO 19115 were presented. This crosswalk is characterized by the complexity of the standards, hierarchically structured and containing a large number of elements. And as an example of across-domain interoperability, the transformation between ISO 19115 and Dublin Core was detailed. It must be mentioned that this second crosswalk is the result of the collaboration in a European project (CEN/ISSS Workshop - Metadata for Multimedia Information - Dublin Core) whose one of their objectives was to create a geographic application profile for Dublin Core standard and its mapping to ISO 19115. This project, supported by the European Committee for Standardization (CEN), includes within its deliverables the style sheets (crosswalk implementation) that transform metadata between ISO 19115 XML representation and Dublin Core RDF representation.

Future lines of this process of crosswalk construction should be oriented to the design of a CASE tool assisting this process. Despite having described a semi-formal method, this process is still error-prone if not done with enough thoroughness. Thus, perhaps the main challenges of this CASE tool will be: provide help in the harmonized description of standards; facilitate the semantic mapping between the source and target standards; and the further automation in the creation of stylesheets. Firstly, XML-Schemas and DTDs could be used to generate as much as possible harmonized descriptions of the source and target standards. Secondly, an initial semantic mapping could be automatically proposed by means of the linguistic analysis of element terms using dictionaries and lexical ontologies. That is to say, the element terms from both standards would be disambiguated against an upper-level ontology in order to recognize possible links. Obviously this linguistic mapping will not be exact but it will probably detect some obvious mappings that can save time of the user. And finally, the automatic creation of XSL documents could be improved. Section 3.3.4 has presented a solution to generate automatically an initial version of the stylesheets. However, most of the additional transformation rules must be still hand-coded. Some of these conversion problems have been already researched by other works. For instance, the problem of maintaining foreign key constraints in the target standard has been studied in (Popa et al., 2002; Fagin et al., 2003). These works describe a research project called Clio, which has obtained optimistic results with the semi-automatic transformation of XML and relational schemas. Nevertheless, other conversion problems are more context-specific and it is difficult to find general patterns applicable in different crosswalks. Thus, further research must be done in the categorization and specification of these rules to facilitate their automatic translation into a series of XSLT instructions.

4

The use of disambiguated thesauri to improve information retrieval

4.1 Introduction

This chapter deals with the use of appropriate vocabularies and classifications within metadata records as an essential way to homogenize metadata content and improve the performance of information retrieval mechanisms in metadata catalogs. This problem is not particular of spatial data infrastructures (the application context of this book). On the opposite, the improvement of information retrieval performance is perhaps the main issue in digital libraries research. Advances in the digital libraries field are directly applicable to the metadata catalogs integrated within a spatial data infrastructure. Therefore, from now on the concepts presented in this chapter will be framed in the context of digital libraries.

As opposite to the largely unstructured information available on the Web, information in digital libraries is explicitly organized, described, and managed. In order to facilitate discovery and access, digital library systems summarize the content of their data resources into metadata records, which can be either introduced manually or automatically generated (e.g.: index terms automatically extracted from a collection of documents; etc.). The focus of this chapter is digital libraries working with metadata records using an agreed metadata schema. Indeed, most digital libraries use structured metadata in accordance with recognized standards such as MARC21 (U.S. Library of Congress, 2004b) or Dublin Core (DCMI, 2004). Moreover, in order to provide accurate metadata, metadata creators use specialized thesauri to fill the content of typical keyword sections. According to ISO 2788 (norm for monolingual thesauri), a thesaurus is a set of terms that describe the vocabulary of a controlled indexing language, formally organized so that the a priori relationships between concepts (for example synonymous terms, broader terms, narrower terms and related terms) are made explicit. Thesauri provide a specialized vocabulary for the homogeneous classification of resources and for supplying users with a suitable vocabulary for the retrieval. For instance, Dublin Core includes a *subject* element (see figure 4.1) and recommends the use of several thesauri like the "Library of Congress Subject Headings" (U.S. Library of Congress, 2004a). And the ISO 19115 geographic metadata standard provides the elements *topicCategory* and *descriptiveKeywords* to include the category (a value from a restricted code list) and the set of keywords which better describe a geographic resource (see figure 4.2).

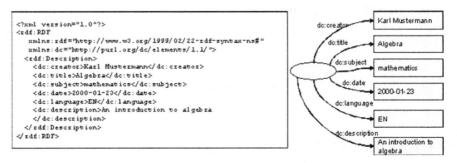

Fig. 4.1. Example of Dublin Core metadata: encoded in RDF (left) and re-expressed using a hedgehog graph

Works like (Swoboda et al., 1999; Clark et al., 2000) present systems where thesauri are used as the basis for discovery services. The first work (Swoboda et al., 1999) is a typical example of a catalog that offers a thesaurus navigator as a retrieval tool. It presents an environmental data catalog (Umweltdatenkatalog, UDK) developed by the German and Austrian authorities. The thesaurus navigator enables the user to browse along a structured set of environmental terms. Once the user has selected a term, all the records linked to this category are presented to the user. The second work (Clark et al., 2000) is more sophisticated and discovers resources that may not be directly linked to the category/term specified by the user query. This second system aims at identifying human experts in different subjects of an application domain. For that purpose, a concept index was built manually and experts were associated with these concepts. After the user specifies a set of concepts, the system searches for experts who either know about one of those concepts or know about concepts "closely" related to "the user's concepts of interest". That is to say, the system evaluates the semantic relatedness using the network representation of the thesaurus. The hits returned are ranked according to the distance between query concepts and the concepts assigned to each expert.

However, if a digital library aims at providing access to the general public (not only constrained to the community of experts that created the resources in the digital library), it is not reasonable to assume that casual users will use the same query terms as the keywords used in metadata records. This discordance between query terms and metadata keywords is even worse in the case of digital libraries handling resources from different application domains, where metadata creators have probably used different thesauri (increasing the

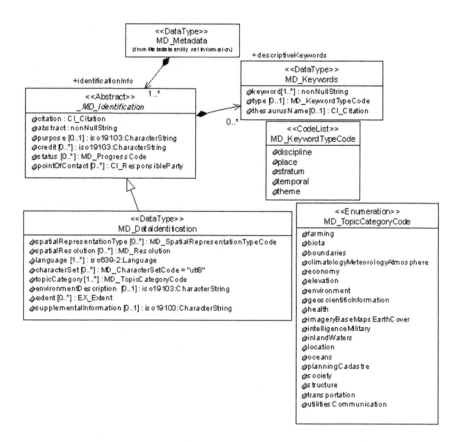

Fig. 4.2. Elements *topicCategory* and *descriptiveKeywords* in ISO 19115

heterogeneity of keywords). This situation implies that discovery in digital libraries cannot be implemented as a simple word matching between the user queries and metadata records. On the contrary, a digital library should be able to understand the sense of the user's vocabulary and to link these meanings to the underlying concepts expressed by metadata records (Bañares et al., 2000).

In order to fill the semantic gap between user queries and metadata records, this work proposes a method for the semantic disambiguation of thesauri with respect to an upper-level ontology, which is closer to the user expressions. Concepts contained in user queries are usually extracted by means of natural language processing techniques (beyond the scope of this chapter) that also make use of similar upper-level ontologies. Therefore, it seems reasonable to use the semantic disambiguation of thesauri as a mechanism that harmonizes concepts in metadata records and user queries. In particular, our method provides the disambiguation against WordNet (Miller, 1990; Fellbaum, 1998), a large-scale lexical database developed from a global point of view that can provide a good kernel to unify, at least, the broader concepts included in distinct thesauri. Our method can be classified as an unsupervised disambiguation method that applies a heuristic voting algorithm and makes profit of the hierarchical structure of both WordNet and the thesauri. Whereas thesaurus hierarchical structure provides the disambiguation context for terms, the hierarchical structure of WordNet enables the comparison of senses from two related thesaurus terms. The heuristic disambiguation method presented in section 4.3 evolves from the initial ideas presented in (Mata et al., 2001), whose first aim was to find a way of interconnecting thesauri from different nature. But in contrast to this preliminary work, this chapter provides a formalized definition of the method; it introduces a more sensible adjusting of the heuristics that were initially proposed; and finally, it includes a more rigorous verification of the quality of the method.

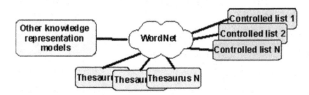

Fig. 4.3. Using WordNet as a unifying system

This disambiguation facilitates a unifying system (see figure 4.3) to express user queries and metadata records but it does not constitute itself the final objective. As mentioned in (Resnik and Yarowsky, 1997), the disambiguation word senses is an intermediate capability that is believed (but not yet proven) to improve natural language applications like machine translation, speech synthesis or information retrieval. In particular, the purpose of this chapter is to integrate this disambiguation within an Information Retrieval System (IRS). In fact, the indexing with WordNet synsets is not new in the context of general text retrieval, some related works can be found in (Gonzalo et al., 1998a; Sanderson, 1994; Voorhees, 1993b; Krovetz and Croft, 1992). In general, the conclusion of these works is that WordNet indexing can improve performance whenever the disambiguation accuracy rate is high (in some cases not less than 90%). These conclusions are probably not extensible to the IRS proposed in this chapter because they were indexing free text and this IRS is constrained to the keywords section of metadata. However, it is expected that the disambiguation accuracy in our IRS will be very high. The first reason is that we are disambiguating the own keywords. As opposed to free text retrieval, we are not going to extract concepts from words that are not essential to the document meaning. Additionally the thesaurus hierarchy provides an accurate and limited context for the disambiguation.

After presenting the semantic disambiguation method, this chapter will present the applicability of this method within an information retrieval system. The vector-space retrieval model (Salton and Lesk, 1968; Salton, 1971) will be adapted to the context of metadata catalogs. Other classical models, like the probabilistic or neural-net based models, would probably perform better in more heterogeneous contexts. However, the initial hypothesis was that in this context, where metadata records are the summary of the desired resource, a simple model may provide satisfactory results. The indexing technique makes profit of this keywords section, whose content has been strategically filled in by selecting terms from disambiguated thesauri. And thanks to the disambiguation, both metadata records and user queries can be homogenously represented as a collection of WordNet synsets (set of synonyms used to express a concept in WordNet), thus enabling the computing of a similarity value, which ranks the results returned by the digital library. Additionally some of the results from the initial experiments of the retrieval system (tested against a geographic data catalog) are presented.

The rest of the chapter is organized as follows. The following section introduces some preliminaries about thesaurus and WordNet. Then, the thesaurus disambiguation method is presented. Section 4.4 details the information retrieval system with the adaptation of the indexing technique to the specific features of metadata schemas. And finally, this chapter ends with some conclusions and future lines.

4.2 Basic concepts about thesaurus and WordNet

4.2.1 Thesaurus

Thesauri are structured, controlled vocabularies of words and phrases that represent conceptual categories (Janée et al., 2003). The formal definition of thesauri is specified by American and international standards. On one hand the American norm ANSI/NISO Z39.19 (ANSI, 1993) establishes guidelines for the construction, format, and management of monolingual thesauri. And on the other hand, the ISO Technical Committee 46, whose remit is Information and Documentation, has delivered two standards defining the norms for monolingual and multilingual thesauri, ISO 2788 (ISO, 1986) (equivalent to Z39.19) and ISO 5964 (ISO, 1985) respectively.

According to these standards, a thesaurus is organized in a set of terms and a set of standardized reciprocal relations on those terms. A term is a word or phrase that represents a conceptual category. Additionally, a term may have an associated human-readable description, or scope note, which defines the concept represented by the term and indicates the term's intended usage. There are two varieties of terms, preferred (or valid) and nonpreferred (or invalid or lead-in). Preferred terms participate in all the relations described below; nonpreferred terms participate in the equivalence relations only. Figure 4.4 displays an overview of relations that are explained below.

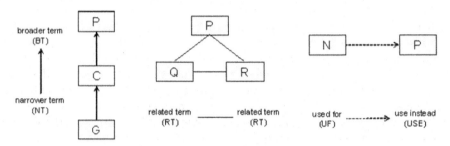

Fig. 4.4. Thesaurus relations

A pair of reciprocal hierarchical relations is the primary means by which thesauri are structured. The narrower (NT) relation relates a preferred term P to another preferred term C that is in some sense a subset of P: as suggested by Z39.19, the concept represented by C may be more specific than that of P, or C may be a component of the whole represented by P, or C may be an instance of the general class represented by P. The narrower relation must be non-reflexive (a term must not be narrower than itself), non-symmetric (two terms must not be mutually narrower than each other), and non-transitive (the narrower relation is logically transitive, that is, if G is narrower than C and C is in turn narrower than P then G is logically narrower than P, but transitive closures must not be reflected in the protocol; rather, they must be left to the client to deduce from first-order relations). The broader (BT) relation is the reciprocal of the narrower relation. A preferred term may be related to any number of broader and narrower terms. The directed graph induced by the narrower relation (equivalently, the broader relation) must be acyclic.

The related (RT) relation relates a preferred term P to another preferred term Q that in some sense intersects P: the concepts represented by P and Q may overlap, or P and Q may be suggestive of each other. The relation must be non-reflexive (a term must not be related to itself), symmetric (if P is related to Q then Q must be related to P), and transitive (if P is related to Q and Q is in turn related to R, then P must be related to R). A preferred term may be related by the related relation to any number of other preferred terms.

A pair of reciprocal equivalence relations ties equivalent terms together. The use-instead (USE) relation maps a nonpreferred term N to a preferred term P that is equivalent to N and that has been designated by the thesaurus as the preferred or canonical term to use in place of N. The used-for (UF) relation is the reciprocal relation that maps P to N. Every nonpreferred term N must be related to at least one preferred term; if more than one, the entire set of N's relations can optionally be designated as a conjunction if N is equivalent to the logical conjunction of the preferred terms.

4.2.2 WordNet

WordNet (Miller, 1990; Fellbaum, 1998) is a public domain electronic lexical database which is considered to be the one of the most important resources available to researchers in computational linguistics, text analysis, and many related areas. It was developed manually at the beginning of the 1980s by George A. Miller and his colleagues at the Cognitive Science Laboratory at Princeton University. Originating from a project whose goal was to produce a dictionary that could be searched conceptually instead of only alphabetically, WordNet has evolved into a system that reflects current psycholinguistic theories about how humans organize their lexical memories.

The basic object in WordNet is a set of strict synonyms called a synset, which represents one underlying lexicalized concept. By definition, each synset in which a word appears is a different sense of that word. There are four main divisions in WordNet: nouns, verbs, adjectives, and adverbs. Within a division, synsets are configured as a semantic net where basic semantic relations (synonymy, hyponymy, meronymy, etc.) are established between them. WordNet can be considered as a large-scale taxonomic class hierarchy, where each word sense corresponds to a taxonomic class in the hierarchy.

As an example, table 4.1 shows the eight different senses of the noun *state*. For nouns the lexical relations include antonymy, hypernymy/hyponymy (*is-a* relation) and three different meronym/holonym (*part-of*) relations. The *is-a* relation is the dominant relation, and organizes the synsets into a set of approximately ten hierarchies. Figure 4.5 shows the *is-a* hierarchy relating the eight different senses of the noun *state*. The synsets with the double border are the actual senses of *state*, and the remaining synsets are either ancestors or descendants of one of the senses. The synsets *group*, *entity*, *state* and *psychological feature* and *phenomenon* in the figure are examples of heads of hierarchies.

WordNet 1.6, the version of WordNet used in the later experiments shown in this chapter, contains more than 99,000 synonym sets and about 120,000 strings (single or compound terms). Table 4.2 gives statistics about WordNet content. According to this, the average number of senses per word is close to one. These figures seem to suggest that polysemy and synonymy occur too infrequently to be a problem for retrieval, but they are misleading. The more frequently a word is used, the more polysemous it tends to be (Zipf, 1945). As mentioned in (Voorhees, 1993b), it is precisely those nouns that actually get used in documents and query statements that are most likely to have many senses and synonyms.

Table 4.1. Senses of *state*

Synset Nr	Synset is-a hierarchy	Definition
6060831	group→social group→organization→unit→ administrative unit→division→department→ government department→federal department→executive department→Department of State	the federal department that sets and maintains foreign policies; "the Department of State was created in 1789"
6074189	group→social group→organization→unit→ political unit→state	a politically organized body of people under a single government; "the state has elected a new president"
6079469	group→social group→organization→polity →government→state	the group of people comprising the government of a sovereign state; "the state has lowered its income tax"
6299747	entity→object→location→region →district→administrative district→country	the territory occupied by a nation; "he returned to the land of his birth"; "he visited several European countries"
6374245	entity→object→location→region →district→administrative district→state	the territory occupied by one of the constituent administrative districts of a nation; "his state is in the deep south"
16185	state	the way something is with respect to its main attributes; "the current state of knowledge"; "his state of health"; "in a weak financial state"
10077290	psychological feature→feeling→emotion→ emotional state→state	(informal) a state of depression or agitation; "he was in such a state you just couldn't reason with him"
10386919	phenomenon→natural phenomenon→chemical phenomenon→state of matter	(chemistry) the three traditional states of matter are solids (fixed shape and volume) and liquids (fixed volume and shaped by the container) and gases (filling the container); "the solid state of water is called ice"

Table 4.2. WordNet content

Term Type	Nr. of terms	Nr. of synsets	Nr. of senses	Mean Nr of senses per term
noun	94474	66025	116317	1.23
verb	10319	12127	22066	2.13
adjective	20170	17915	29881	1.48
adverb	4546	3575	5677	1.24
total	**121962**	**99642**	**173941**	

Finally, it must be mentioned that numerous interfaces have been written in different languages to access the files where WordNet lexical concepts and relationships are stored. Apart from the original C libraries, links to interfaces written in other languages can be found through the WordNet Web site[1].

4.3 The Semantic Disambiguation of Thesauri

4.3.1 State of the art in Semantic Disambiguation

Word Sense Disambiguation (WSD) is perhaps the greatest existing problem at the lexical level in natural language processing (Resnik and Yarowsky,

---

[1] http://www.cogsci.princeton.edu/~wn/links.shtml

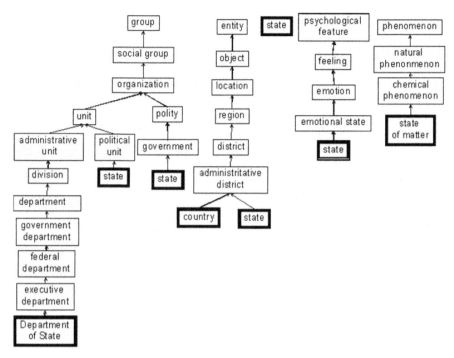

Fig. 4.5. The *is-a* hierarchy for eight different senses of the noun *state*

1997), and this skill is applicable to tasks such as machine translation, speech synthesis and information retrieval. A word is polysemic if its sense changes depending on the context. The problem of disambiguation consists in determining which one of the senses of an ambiguous word is invoked in a particular context composed of a set of words related to the ambiguous word.

The earliest word sense disambiguation methods used hand-coding of knowledge to disambiguate word senses. In these systems, each word to be disambiguated would need to be hand-tagged with the correct piece of information (e.g. part-of-speech, sense, etc.) which would be useful in the disambiguation process. Therefore, it was difficult to obtain a comprehensive set of the necessary disambiguation knowledge and even more difficult to manually maintain and further expand the disambiguation knowledge to handle real world sentences.

In order to solve this problem, researchers started to essay the use of either machine readable dictionaries or machine learning techniques. Nowadays all approaches to the word sense disambiguation problem can be categorized in three general strategies (Sanfilippo et al., 1999): knowledge-based disambiguation, which uses an explicit lexicon; corpus-based disambiguation, where the information about word senses is gathered from training on a large corpus; or a third alternative called the hybrid approach, which combines aspects of the previous methodologies. A revision of works related to each of these categories is presented in next subsections.

Additionally, it must be mentioned that the performance of WSD algorithms is usually measured by comparison with a baseline called the most frequent heuristic (Gale et al., 1992). This baseline consists in taking always the most frequent sense of a word. In order to be of any value, a WSD algorithm should perform at least as well as the "most frequent" heuristic. For instance, this baseline might be implemented by making use of WordNet. WordNet sorts the senses (synsets) of each word starting from the most to the least frequent, and it always starts with the noun senses of words. Therefore, the performance of the "most frequent" heuristic could easily be evaluated by assigning each ambiguous content word the sense of the first synonym set that appears in its noun group. The most frequent heuristic usually has a success rate of around 50% with variations depending on the texts being disambiguated. In general there is still a lack of accurate WSD methods. As mentioned in (Gonzalo et al., 1998b), experiments with unrestricted WSD on large corpora show that the best algorithms do not outperform the "most frequent" heuristic.

Knowledge-based

Under this approach disambiguation is carried out using information from an explicit lexicon or knowledge base. The lexicon may be a machine readable dictionary, thesaurus or it may be hand-crafted. This is one of the most popular approaches to word sense disambiguation and amongst others, work has been done using existing lexical knowledge sources such as WordNet, the Longman Dictionary of Contemporary English (LDOCE) (LONGMAN, 2003) or the Roget's International Thesaurus (Chapman, 1992).

One of the earliest works is the one presented in (Lesk, 1986). Lesk starts from the idea that a word's dictionary definitions are likely to be good indicators for the senses they define. Given an ambiguous word and a set of words occurring in a particular context of the ambiguous word, it is computed for each sense of the ambiguous word the number of words shared by the context and its dictionary definition. Finally, the sense with the highest overlap is chosen.

(Sussna, 1993) presents a method that enables document indexing using a massive semantic network, WordNet. This method relies on the use of the noun taxonomy of WordNet and the notion of conceptual distance among concepts. Conceptual distance tries to provide a basis for determining closeness in meaning among pairs of words, taking as reference a structured hierarchical net. Conceptual distance between two concepts is defined in (Rada et al., 1989) as the length of the shortest path that connects the concepts in a hierarchical semantic net. In this method defined by Sussna, input terms with multiple senses have been disambiguated by finding the combination of senses from a set of contiguous terms (context window) which minimizes total pairwise distance between senses.

(Agirre and Rigau, 1996) presents a method that also makes use of Word-Net. It is based on an elaboration of the conceptual distance: the conceptual density. According to Agirre and Rigau, the measure of conceptual distance among concepts should be sensitive to:

- The length of the shortest path that connects the concepts involved.
- The depth in the hierarchy: concepts in a deeper part of the hierarchy should be ranked closer.
- The density of concepts in the hierarchy: concepts in a dense part of the hierarchy are relatively closer than those in a sparser region.
- The measure should be independent of the number of concepts we are measuring.

The conceptual density (CD) formula that is defined in (Agirre and Rigau, 1996) compares areas of subhierarchies in WordNet, where each subhierarchy represents a sense of the word to disambiguate. Within each subhierarchy, the senses of the context words are also taken into account. Then, given a concept c at the top of a subhierarchy and given $nhyp$ (mean number of hyponyms per node), the conceptual density for c when its subhierarchy contains m marks (senses of either the word to disambiguate or the words in the context) is computed as the division between the expected area of a subhierarchy containing m marks and the real area (number of descendants of c). Additionally, it can be observed in equation 4.1 that the formula includes the parameter 0.20, which was computed experimentally. Finally, the subhierarchy with the highest density will correspond to the disambiguated sense.

$$CD(c,m) = \frac{\sum_{i=0}^{m-1} nhyp^{i^{0.20}}}{descendants_c} \tag{4.1}$$

Corpus-based

This approach attempts to disambiguate words using information which is gained by training on some corpus, rather than taking it directly from an explicit knowledge source. The means used to assign senses to ambiguous words are then distributional information and context words. Distributional information about a word is simply its frequency. Context words are the words found to the right and/or the left of a certain word, thus collocational information.

There are two possible approaches to corpus-based WSD systems: supervised and unsupervised WSD. In supervised methods the training is carried out on a disambiguated corpus, where the sense of each polysemous lexical item has been previously labeled. During training on a disambiguated corpus probabilistic information about context words as well as distributional information about the different senses of an ambiguous word are collected. In the testing phase, the sense with the highest probability computed on the basis of the training data (context words) is chosen. Unsupervised methods, on the other hand, are applied to a raw text material, where no previous sense annotation has been performed. Sometimes, even knowledge-based disambiguation methods are cited as unsupervised methods because they do not need sense tagged corpora.

An example of supervised methods is the Bayesian classifier introduced by (Gale et al., 1992). The essence of the method is to compute the probability of each sense s_i of an ambiguous word given the context C and to choose the most probable sense. This probability, $P(s_i|C)$, is computed using the Bayes' Theorem:

$$P(s_i|C) = \frac{P(C|s_i)P(s_i)}{P(C)} \tag{4.2}$$

If we only wish to maximize this quantity, the denominator can be ignored, and if we assume the independence hypothesis between the words in a context, that are clearly not independent of each other, we can factorize the computation of:

$$P(C|s_i) = \prod_{w \in C} P(w|s_i) \tag{4.3}$$

where w denotes a word in the context.

$P(w|s_i)$ and $P(s_i)$ are computed via Maximum-Likelihood estimation:

$$P(w|s_i) = \frac{N(w, s_i)}{N(s_i)} \tag{4.4}$$

$$P(s_i) = \frac{N(s_i)}{N(a)} \tag{4.5}$$

where $N(w, s_i)$ is the number of occurrences of w in a context of sense s_i, $N(s_i)$ is the number of occurrences of s_i in the training corpus, and $N(a)$ is the total number of occurrences of the ambiguous word a.

Although supervised methods can get good results, all of them need labeled texts for training the algorithms. This text collection is not available for many domains and creating it can be too expensive. In these situations, unsupervised methods may provide a good alternative as they try to distinguish among the senses of a polysemic word using only the features that can be automatically extracted from unlabeled texts. Strictly speaking, completely unsupervised disambiguation is not possible. The mere fact of labeling a word as belonging to one sense or another requires some characterization of the sense to be provided. However, we can discriminate among different classes in which each occurrence of the ambiguous word has some different characteristics and consider these classes as possible senses of the word.

An example of unsupervised disambiguation is the dynamic matching technique presented by (Radford et al., 1995), which examines all instances of a

given term in a corpus and compares the contexts in which they occur for
common words and syntactic patterns. A similarity matrix is thus formed
which is subject to cluster analysis to determine groups of semantically related instances of terms.

Hybrid

These methods can be neither properly classified as knowledge nor corpus
based because they combine part of those approaches. They merge the definition of senses in dictionaries and other lexical resources with training information extracted by supervised or unsupervised methods.

For instance, (Yarowsky, 1992) describes a method using statistical models
of the major Roget's Thesaurus categories. Roget's categories (1042 categories
in the 1997 version) serve as approximations of conceptual classes. He defines
the senses of a word as the categories listed for that word in the Roget's
Thesaurus. Sense disambiguation will constitute selecting the most probable
listed category given the context. The process consists of three steps:

1. The goal of the first step is to collect a set of words that are typically
   found in the context of a Roget's category. To do this, the method extracts concordances of 100 surrounding words for each occurrence of each
   member of the category in the training corpus used (an electronic version
   of Grolier's Encyclopaedia).
2. Identify salient words in the collective context, and determine weights. A
   salient word is one which appears significantly more often in the context
   of a category than at other points in the corpus, and hence is a better
   than average indicator for the category. With $\frac{P(w|RCat)}{P(w)}$ it is formalized
   the probability of a word appearing in the context of a Roget's category
   divided by its overall probability in the corpus.
3. Use the resulting weights to predict the appropriate category for a polysemous word occurring in novel text. When any of the salient words derived
   in step 2 appear in the context of an ambiguous word, there is evidence
   that the word belongs to the indicated category. If several such words appear, the evidence is compounded. Using Bayes' rules, the method sums
   their weights, over all the words in the context, and determines the category for which the sum is greatest:

$$ARGMAX_{RCat} \sum_{w\ in\ C} log\frac{P(w|RCat)P(RCat)}{P(w)}$$

(Resnik, 1995a) presents a method for automatic sense disambiguation of
nouns appearing within sets of related nouns, the kind of data one finds in on-
line thesauri. Disambiguation is performed with respect to WordNet senses. He
makes the assumption that word groupings have been obtained through some
black box procedure, an on-line thesaurus or an unsupervised word clustering
method. Each word group is considered the context for each word contained

in it. The disambiguation algorithm is inspired by the observation that when two polysemic words are similar, their most informative subsumer provides information about which sense of each word is the relevant one (Resnik, 1995b). Given two words w_1 and w_2, the most informative subsumer of both words is the concept c that maximizes their semantic similarity. The computation of the semantic similarity uses the WordNet *is-a* taxonomy for nouns and is calculated as

$$sim(w1, w2) = max_{c \in subsumers(w_1, w_2)}[-log \ Pr(c)] \qquad (4.6)$$

where $subsumers(w_1, w_2)$ is the set of WordNet synsets that are ancestors of both w_1 and w_2, in any sense of either word. Probability estimates $(Pr(c))$ are derived from a corpus by computing the number of nouns having a sense subsumed by the concept c divided by the total number of noun instances observed.

4.3.2 Description of the semantic disambiguation method

This section presents an knowledge-based disambiguation method based on the hierarchical structure of WordNet, which is similar to the methods described in (Sussna, 1993; Agirre and Rigau, 1996; Resnik, 1995a). But unlike the work of (Resnik, 1995a) it does not imply a training corpus to estimate probabilities for calculating the semantic similarity. On the contrary, it takes advantage of the fact that thesauri have a hierarchical structure, which may serve as the words context to evaluate a particular term.

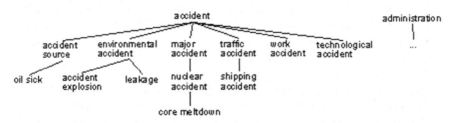

Fig. 4.6. The branch *accident* of the GEMET thesaurus

This method considers a thesaurus as a set of branches, similar to trees, whose nodes are the terms that maintain associations with their broader (ascendants) or narrower terms (descendants). Each branch corresponds to a tree whose root is a term with no broader terms in the thesaurus. The objective of the method is to analyze the thesaurus terms and, for each word in the thesaurus, determine the "closest" sense to the senses of the rest of the words in the whole branch. Figure 4.6 shows an example of a thesaurus that contains a branch whose first root is the term *accident* and all its descendants; the second one is the term *administration* with its descendants; and so on.

For each term belonging to a branch, the disambiguation method assumes that other terms in the branch constitute its context. Therefore, as the branch is traversed, all possible senses of each term in the branch are extracted from WordNet. In case a term is a compound term (more than one word) and is not included in WordNet, the senses for each word are extracted. A sense or concept in WordNet is represented with a synset (a set of synonyms, represented by a number) and as synsets in WordNet also maintain a hierarchical structure, it is possible to obtain a synset path for each extracted sense. For example, the term *accident* has two WordNet senses, and therefore, two corresponding synset paths can be derived:

- synset 1 : *[5443572] accident*
  - synset path 1: *event→happening→trouble→misfortune→mishap→accident*
- synset 2 : *[5443380] accident, fortuity, chance, event*
  - synset path 2: *event→happening→accident*

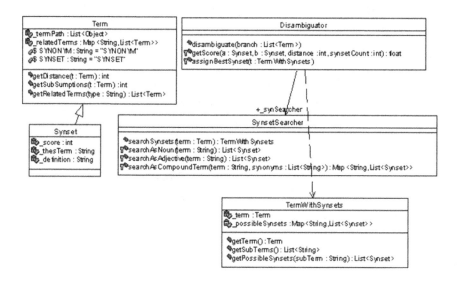

Fig. 4.7. Classes to implement the disambiguation method

Instead of working on the idea of the closest sense by means of probability theory as in (Gale et al., 1992), it was chosen a voting system that integrates the words in the context without assuming the independence hypothesis often used in Bayesian classifiers for simplifying the complexity of the operations. This method uses the hierarchical structure of WordNet on the assumption that: "the more similar two senses are, the more hypernyms they share". Given a synset path (i.e. a possible sense) of a term, the voting system compares it with the rest of synset of the other terms in the same branch (i.e. the context). Additionally, in the case of having a compound term, a synset path of a subterm would also vote for the synset paths associated with the rest of subterms of this compound term. For each pair of synset paths, the system counts the number of hypernyms (WordNet synsets) that subsume both of them, giving an accumulated result for the initial synset path. And once the results are obtained for all the synset paths of a term, the synset path with the higher number of votes is chosen as the disambiguated sense.

```
 1:  public void disambiguate(List branch) {
 2:    // search possible synsets for every term in the branch
 3:    List branchWithSynsets = new LinkedList();
 4:    for (int i=0;i< branch.size();i++) {
 5:      Term term= (Term) branch.get(i);
 6:      TermWithSynsets tws = _synSearcher.searchSynsets(term);
 7:      branchWithSynsets.add(tws);
 8:    }
 9:    // Every possible synset associated with a term votes to the rest of synsets
10:    // associated to the terms in the branch
11:    for (int i=0;i< branchWithSynsets.size();i++) {
12:      TermWithSynsets term1 = (TermWithSynsets) branchWithSynsets.get(i);
13:      List subterms1 = term1.getSubTerms();
14:      for (int m1 = 0; m1 < subterms1.size(); m1++) {
15:        List synsets1 = term1.getPossibleSynsets(subterms1.get(m1));
16:        for (int s1=0; s1< synsets1.size();s1++) {
17:          Synset syn1 = (Synset) synsets1.get(s1);
18:          for (int j=i;j< branchWithSynsets.size();j++) {
19:            TermWithSynsets term2 = (TermWithSynsets) branchWithSynsets.get(j);
20:            int distance= term1.getTerm().getDistance(term2.getTerm());
21:            List subterms2 = term2.getSubTerms();
22:            for (int m2 = 0; m2 < subterms2.size(); m2 ++){
23:              if ((i!=j)||(m1!=m2)) { // a subterm doesnt vote for itself
24:                List synsets2 = term2.getPossibleSynsets(subterms2.get(m2));
25:                for ( int s2 = 0; s2 < synsets2.size(); s2++){
26:                  Synset syn2 = (Synset) synsets2.get(s2);
27:                  //syn1 votes for syn2
28:                  float score2 = getScore(syn1,syn2,distance
29:                    ,synsets1.size()*subterms1.size());
30:                  if (score2>0) syn2.addScore(score2);
31:                  // syn2 votes for syn1
32:                  float score1 = getScore(syn2,syn1,distance
33:                    ,synsets2.size()*subterms2.size());
34:                  if (score1>0) syn1.addScore(score1);
35:                } // for s2
36:              } // if
37:            } // for m2
38:          } // for j
39:        } // for s1
40:      } // for m1
41:      assignBestSynset(term1); // assign best synsets for term1.getTerm()
42:    } // for i
43: }
```

Fig. 4.8. Code of the *disambiguate* method in the *Disambiguator* class

Figure 4.7 shows an object-oriented class diagram which describes the main components involved in this disambiguation algorithm. The *Term* class represents a term contained in a thesaurus branch by means of its path, i.e.

the sequence of broader and narrower terms between the root term of the branch and the intended term. This class also includes methods to obtain the depth of a term in the branch (*getDepth*); to compute the shared hypernyms with another term (*getSubsumptions*); and to obtain the distance between two terms (*getDistance*, see later comments). *Term* objects also have references to related terms (synonyms, disambiguated synsets, and so on). The *Synset* class extends the *Term* class to represent the specifics of WordNet synsets: the WordNet definition; the thesaurus term (attribute *thesTerm*) that may be associated to a synset; and the score (attribute *score*) that is given to a synset so as to reflect the closeness to a thesaurus term. The *Disambiguator* class is the one responsible for the disambiguation of terms in a thesaurus branch. The *disambiguate* method of this class receives the list of terms in a branch and performs the disambiguation. Figure 4.8 shows the code (in Java programming language) of this method which has two separate parts: firstly, the method tries to find the possible synsets that may be associated with a term (lines 3-8); and secondly, the voting among the different synset paths is applied (lines 9-43).

```
1:   public TermWithSynsets searchSynsets(Term term){
2:     TermWithSynsets tws = new TermWithSynsets(term);
3:     String singleTerm = (String)term.getOriginalTerm();
4:     // search as noun
5:     List synsets = searchAsNoun(singleTerm);
6:     int count = synsets.size(); // count stores the number of found synsets
7:     // if there are no synsets, seach synsets associated with synonyms
8:     if ((term.getSynonyms().size()>0)&&(count == 0))
9:       for (int i=0;(i<term.getSynonyms().size())&&(count==0);i++){
10:        synsets=searchAsNoun((String)term.getSynonyms().get(i));
11:        count = synsets.size();
12:      }
13:    if (count == 0) {
14:      // There are  no synsets associated with term or synonyms
15:      // Separate the term into subterms and look for synsets
16:      Map map=searchAsCompoundTerm(singleTerm,term.getSynonyms());
17:      tws.setPossibleSynsets(map);
18:    } else { // There are synsets associated with term
19:      if ((term.getSynonyms().size()>0)&&(count>1))
20:        // There are more than one synsets and we have synonyms
21:        // We will try to minimize the number of synsets by means of synonyms
22:        synsets=minimizeSynsetsWithSynonyms(term.getSynonyms(),synsets);
23:      tws.setPossibleSynsets(singleTerm,synsets);
24:    }
25:    return tws;
26: }
```

Fig. 4.9. The *searchSynsets* method in the *SynsetSearcher* class

For the first part of the *disambiguate* method, *searchSynsets* method of the *SynsetSearcher* class is used. This method receives a *Term* object and returns an instance of *TermWithSynsets* class. The *TermsWithSynsets* class represents a container of a *Term* object and its possible related synsets in WordNet. The real advantage of this class is that it is also able to manage compound terms. The *subTerms* method returns the words that form part of the term, and *getPosssibleSynsets* returns the synsets that may be associated to a given subterm. The *SynsetSearcher* class acts as a bridge between the disambiguation method implemented in Java and the WordNet native libraries, which are implemented in C and enable the access to WordNet files. The code of the *searchSynsets* method of class *SynsetSearcher* is depicted in figure 4.9. This method tries different strategies to find possible synsets. Firstly, it searches the term without any modifications in WordNet (invocation to the *searchAsNoun* method in line 5). Secondly, it tries to find WordNet synsets connected to synonyms (lines 8-12). Thirdly, if no synsets have been found, it separates the term into subterms and looks up them (invocation to the *searchAsCompoundTerm* method in lines 13-17) into WordNet. This strategy takes into account that a subterm may be an adjective (the *searchAsAdjective* method is used in the implementation of the *searchAsCompoundTerm* method). And finally, this method also considers the reduction of synsets by means of synonyms (lines 18-24).

Regarding the score given by one synset path to another, the initial idea was to assign each other the total number of shared hypernyms (the number of shared hypernyms is computed by method *getSubsumptions*). For instance, the two aforementioned synset paths for the term *accident* would assign each other two votes because they share the synsets *event* and *happening*. Let us observe that they would not receive the third vote by the synset *accident* because the depth is different:

- synset path 1: *event→happening→trouble→misfortune→mishap→accident*
- synset path 2: *event→happening→accident*

However, as a result of the initial tests of the method (see section 4.3.3), three criteria were applied to correct this score. These criteria are slightly related to the aspects used in (Agirre and Rigau, 1996) to define the conceptual distance (the length of a path of concepts in WordNet, the hierarchy and the density depth). In order to facilitate the understanding of these criteria, they will be explained in parallel with the example in table 4.3 and the code shown in figure 4.10. On one hand, table 4.3 shows the scores given by synset paths in the branch *accident* (see figure 4.6) to the synset path *event→happening→trouble→misfortune→mishap→accident* of the term *accident*. The column *sco* shows the final score given by each synset path after applying the three criteria. And the total score for the voted synset is marked on the right of this synset path. On the other hand, figure 4.10 displays the *getScore* method of the *Disambiguator* class, which is called in lines 28 and 32 of figure 4.8.

1. Firstly, lower level WordNet concepts (synsets) have longer paths and then, share more sub-hierarchies. Therefore, the number of shared hypernyms (*sub* column in table 4.3) is divided by the length of the path, i.e. the depth of the WordNet concept. For instance, synset path

Table 4.3. Voting for synset path *event→happening→trouble→misfortune→mishap →accident* of term accident

Term	Subterm	Synset path	sub	dep	dis	pol	sco
accident							
		event→happening→trouble→misfortune→mishap →accident			**total score = 3.143**		
		event→happening→accident			*it doesn't vote*		
accident→accident source							
	accident	event→happening→trouble→misfortune→mishap →accident	6	6	1	4	0.250
		event→happening→accident	2	3	1	4	0.167
	source	*7 synsets without subsumers*					
accident→accident source→oil slick							
		entity→object→film→oil_slick	0	4	2	1	0.000
accident→environmental accident							
	accident	event→happening→trouble→misfortune→mishap →accident	6	6	1	4	0.250
		event→happening→accident	2	3	1	4	0.167
	environmental	*2 synsets without subsumers*					
accident→environmental accident→explosion							
		event→happening→discharge→explosion	2	4	2	3	0.083
		act→action→change→change_of_integrity→explosion	0	5	2	3	0.000
		act→action→change→change_of_state→termination →release→plosion	0	7	2	3	0.000
accident→environmental accident→leakage							
		event→happening→movement→change_of_location →flow→discharge→escape	2	7	2	1	0.143
accident→major accident							
	accident	event→happening→trouble→misfortune→mishap →accident	6	6	1	4	0.250
		event→happening→accident	2	3	1	4	0.167
	major	*1 synset without subsumers*					
accident→major accident→nuclear accident							
	accident	event→happening→trouble→misfortune→mishap →accident	6	6	2	4	0.125
		event→happening→accident	2	3	2	4	0.083
	nuclear	*2 synsets without subsumers*					
accident→major accident→nuclear accident→core meltdown							
	core	*8 synsets without subsumers*					
	meltdown	*no synsets in WordNet*					
accident→traffic accident							
	accident	event→happening→trouble→misfortune→mishap →accident	6	6	1	4	0.250
		event→happening→accident	2	3	1	4	0.167
	traffic	*3 synsets without subsumers*					
accident→traffic accident→shipping accident							
	accident	event→happening→trouble→misfortune→mishap →accident	6	6	2	4	0.125
		event→happening→accident	2	3	2	4	0.083
	shipping	*2 synsets without subsumers*					
accident→work accident							
	accident	event→happening→trouble→misfortune→mishap →accident	6	6	1	4	0.250
		event→happening→accident	2	3	1	4	0.167
	work	*7 synsets without subsumers*					
accident→technological accident							
	accident	event→happening→trouble→misfortune→mishap →accident	6	6	1	4	0.250
		event→happening→accident	2	3	1	4	0.167
	technological	*2 synsets without subsumers*					

$event \rightarrow happening \rightarrow trouble \rightarrow misfortune \rightarrow mishap \rightarrow accident$ (depth=6) is likely to receive more votes than synset path $event \rightarrow happening \rightarrow accident$ (depth=3) if this restriction is not applied. In table 4.3, the depth of every synset path is shown in column *dep*. This criterion is applied in line 5 of figure 4.10.

2. Secondly, not all the terms in the context should be valued in the same way. The number of votes provided by the synset paths of a term A to a synset path of a term B are divided by the distance between the two terms (A and B) in the thesaurus. For instance, obtaining the scores for the synsets of the term *accident*, the term *environmental accident* is more important than the term *explosion* because it is closer in the hierarchy. Figure 4.11 shows the *getDistance* method of the *Term* class. This method returns the length of the path between the two terms in the thesaurus hierarchy of BT/NT terms, equivalent to the semantic distance proposed in (Rada et al., 1989). In table 4.3, the distance of every synset path is shown in *dis* column. This criterion is applied in line 7 of figure 4.10.

3. And thirdly, the most polysemic terms in the context vote more times since each one of their senses has the opportunity to vote. The number of votes provided by a synset path is divided by the number of senses of the term to which it belongs. For instance, term *accident source* votes with its nine synset paths, meanwhile term *leakage* only votes with one synset path. In table 4.3, the polysemic value of every synset path is shown in *pol* column. The disambiguate method (line 28 and 32 of figure 4.8) computes this polysemic value as the product of the number of subterms and the number of possible synsets for each subterm. This criterion is applied in line 9 of figure 4.10.

```
1:   private float getScore(Synset a, Synset b, int distance, int synsetCount) {
2:      //initial value without applying factors
3:      float res = a.getSubsumptions(b);
4:      // factor concerning synset depth
5:      if (a.getDepth()>1)) res=res/a.getDepth();
6:      // factor concerning distance in the branch
7:      if (distance>1) res/=distance;
8:      // factor concerning polisemy
9:      if (synsetCount >1) res/=synsetCount;
10:     return res;
11: }
```

Fig. 4.10. The *getScore* method in the *Disambiguator* class

4.3.3 Testing the method

As mentioned in (Uzuner, 1998), the evaluation of WSD algorithms is a more or less subjective process. This is partly due to the lexical resources, like WordNet and LDOCE, which concentrate their efforts in completeness and

```
1:  public int getDistance(Term b) {
2:    int tam1=getTermPath().size();
3:    int tam2=b.getTermPath().size();
4:    // count stores the size of the closest subsumer from a and b
5:    int count=getSubsumptions(b);
6:    // It returns the length of the path from term a to the closest
7:    // subsumer and from the closest subsumer to b */
8:    return(tam1-count+tam2-count);
9:  }
```

Fig. 4.11. The *getDistance* method in the *Term*class

thus make very subtle distinctions between word senses. This sometimes makes it very difficult for even human beings to distinguish between two senses of the same word. Also, many times two people might disagree on the best sense of a word that would fit into a context. Thus, in some cases, it is almost impossible for a WSD algorithm to distinguish between two senses of a word. Although we believe that it is unnecessary to make such subtle distinctions between senses, we stick to WordNet senses of the words in evaluating the performance of our system.

In the case of free text word sense disambiguation, there are some datasets prepared for testing WSD algorithms. One of the most commonly used is the SemCor (Miller et al., 1993) corpus which is a subset of the Brown corpus. SemCor has a total of 186 files (each file containing an average of 800 sentences) where each word in a sentence has been tagged with its correct part-of-speech and sense number taken from WordNet. However, to the best of our knowledge, there is not such a test-bed for thesauri disambiguation. Therefore, we selected on our own criteria a thesaurus to verify the viability of the disambiguation method. The thesaurus selected was GEMET (GEneral Multilingual Environmental Thesaurus) (EEA, 2001), a thesaurus for the classification of environmental resources has been disambiguated. GEMET was developed by the European Environment Agency and the European Topic Centre on Catalogue of Data Sources together with international experts and contains a core terminology of 5,400 generalized environmental terms and their definitions. The thesaurus is multilingual, with all terms translated into 19 languages: Bulgarian, Czech, Danish, Dutch, English (and American English), Finnish, French, German, Greek, Hungarian, Italian, Norwegian, Portuguese, Russian, Slovak, Slovenian, Spanish, and Swedish.

First of all and before applying the method, it was realized that without splitting the compound terms, only 30% of the terms could be found in WordNet (see table 4.4). As it can be deduced, it was necessary the division of compound terms to increment the potential of the disambiguation algorithm. Furthermore, morphological techniques were used to reduce the number of not-found words and to search adjectives associated with a noun. For instance, *administrative* is associated with *administration*. The statistics of senses per Word are shown in table 4.5. This way, the number of not-found words was reduced to 9.53%, corresponding to verbs and adverbs. Some technical terms

Table 4.4. GEMET terms without split-ting compound terms

Senses	# words	% words
0	3862	69.724
1	993	17.927
2	296	5.344
3	164	2.961
4	85	1.535
5	47	0.849
6	42	0.758
7	18	0.325
8	12	0.217
9	6	0.108
10	5	0.09
11	1	0.018
12	0	0
13	1	0.018
14	0	0
15	3	0.054
16	2	0.036
17	0	0
18	0	0
19	0	0
>=20	2	0.036

Terms: 5539

Top terms: 106

Split terms: 0

Words: 5539

Probability of finding correct sense: 22.371%

Table 4.5. GEMET terms splitting compound terms

Senses	# words	% words
0	956	9.53
1	2562	25.538
2	1835	18.291
3	1480	14.753
4	663	6.609
5	687	6.848
6	677	6.748
7	582	5.801
8	124	1.236
9	181	1.804
10	120	1.196
11	56	0.558
12	14	0.14
13	30	0.299
14	18	0.179
15	17	0.169
16	11	0.11
17	10	0.1
18	0	0
19	1	0.01
>=20	8	0.08

Terms: 5539

Top terms: 106

Split terms: 3601

Words: 10032

Probability of finding correct sense: 45.178 %

were not found still due to the fact that WordNet is a global knowledge base. But in general, it can be concluded that WordNet is a suitable tool as an upper-level ontology. Another fact extracted from table 4.5 is that only 25% of the words searched in WordNet are monosemic. The a-priori probability of finding the correct synset of a word is 45%. Therefore, it made sense to apply the disambiguation algorithm and find the appropriate sense.

As explained before, there is no test-bed for the disambiguation of thesauri and we had to assign manually the correct synset of each term to evaluate later the performance of the method. This task is very subjective and we preferred to restrict our initial tests to one thesaurus branch, the branch *administration* of GEMET thesaurus with 105 terms. Before performing the tests, we implemented the most frequent heuristic and we obtained and accuracy of 63.4% for the branch *administration*. The accuracy for this branch is very high and it may be due to the general terms contained in this branch.

Then, we tested slight variations of our disambiguation algorithm on the branch *administration* of GEMET (a branch containing 105 terms and 194 words splitting compound terms not found in WordNet). These variations were oriented to obtain progressively a better accuracy (see table 4.6 for accuracy details). The tests performed were the following:

Table 4.6. Accuracy obtained with different variations of the disambiguation algorithms

Senses	# words	% words
0	10	5.155
1	49	25.258
2	32	16.495
3	21	10.825
4	21	10.825
5	16	8.247
6	10	5.155
7	15	7.732
8	8	4.124
9	4	2.062
10	6	3.093
11	2	1.031

Algorithm	errors	accuracy %
Most Frequent Heuristic	71	63.40206
Without the three criteria	70	63.91753
With the three criteria	56	71.13402
With the three criteria and using synonyms	54	72.16495

Terms: 105

Split terms: 72

Words: 194

Probability of finding correct sense: 44.58%

Table 4.7. Polysemy of the branch *biosphere*

Senses	# words	% words
0	61	10.627
1	235	40.941
2	96	16.725
3	47	8.188
4	38	6.62
5	32	5.575
6	32	5.575
7	20	3.484
8	2	0.348
9	4	0.697
10	4	0.697
11	0	0
12	0	0
13	2	0.348
14	0	0
15	1	0.174

Terms: 397

Split terms: 160

Words: 574

Probability of finding correct sense: 56.458%

- Firstly, the disambiguation algorithm was tested without applying the three criteria mentioned in the previous section. As a result of this test, around 64% of the words were successfully disambiguated. In this test some expected failures were observed; some failures were due to the fact that the longest paths were most voted; others because of wrong senses that were voted by remote terms in the thesaurus; and others due to the fact that the most polysemic words cast a vote for each one of their senses, many of them erroneous.

- Secondly, we tested the method having into account the three criteria and the results were improved to a 71.13% rate of successful disambiguated terms.

- After these initial results, other information contents in the thesaurus were considered to check whether they could reduce the polisemy of terms. As a first step, we started with synonyms and searched the synsets that might be associated with them in WordNet. In case of finding monosemic synonyms, we chose this synset as the right sense. Otherwise, in case of finding polysemic synonyms, we reduced the polysemy of the original term by intersecting lists of synsets. Although GEMET only includes a few synonyms, this modification achieved an accuracy of 72.16% for the branch *administration*. Other thesauri with more synonyms would have profited more from this modification.

- Additionally, it was also considered the use of related terms and the glosses of the definitions for the reduction of polysemy. But in this case, the method did not yield better, and furthermore, the complexity in time and memory was higher.

Finally, the tests were also extended to other branches of GEMET thesaurus. It was noticed that the best results were obtained for more specific branches. For instance, branches like *biosphere* (see table 4.7 with the characteristics of this branch) obtained an accuracy of 95%. This branch contains a high percentage of monosemic terms and for the rest of the terms the branch provides a rich context for the disambiguation. Obviously, not all the branches will provide such a perfect context for the disambiguation algorithm. But at least, it can be concluded that our disambiguation algorithm improves the performance of the most frequent heuristic.

4.4 The information retrieval model

4.4.1 State of the art in sense based information retrieval

There has been several research works that have applied the disambiguation to an information retrieval system for searching on free-text data. A good revision of such works can be found in (Uzuner, 1998).

(Krovetz and Croft, 1992) examines two test collections to study both the amount of lexical ambiguity in the collections and its effect on retrieval performance. They find that even these relatively small, specialized collections contain words used in multiple senses, but that retrieval effectiveness is not strongly affected by ambiguity, in part because documents with many words in common with a query (and are thus ranked highly with regard to that query) tend to use the words in the same senses as the query. Therefore they concluded that word sense disambiguation did not have a very important impact on information retrieval, but that disambiguation could be beneficial to information retrieval when the collection contained more diverse subject matter and there were a few words in common between the query and the document.

More specifically in the context of WordNet, (Voorhees, 1993b) describes an automatic indexing procedure that uses the "is-a" relations contained within WordNet and the set of nouns contained in a text to select a sense for each polysemous noun in the text. Using this indexer/disambiguator, a document is represented by a vector in which some of the terms correspond to word senses and some correspond to word stems. Then, this indexer was applied to an information retrieval system and a large-scale test was performed against standard test collections. These experiments showed that the performance decreased rather than increasing. The overall degradation of performance was mostly due to the difficulty of disambiguating senses in short query statements. In fact, she showed that trying to disambiguate the query in addition to the corpus made the results worse, especially in cases where the query was very short.

(Sanderson, 1994) presents results which show that the disambiguation process usually affect the performance of the IRS negatively. Apparently, this confirms that query/document matching strategies already perform an implicit disambiguation. Sanderson estimates that in order to be of any practical use and in order to improve the performance of an IRS, a disambiguation algorithm has to work with at least 90% of accuracy. The conclusion of Sanderson from his research is that "word sense ambiguity is only problematic to an IRS when it is retrieving from very short queries. In addition if a word-sense disambiguator is to be of any use to an IRS, then it must be able to resolve word sense to a high degree of accuracy".

The most optimistic work in this area is probably the research done by (Gonzalo et al., 1998a). There, the vector space model for text retrieval was shown to give better results (up to 29% better) if WordNet synsets were chosen as the indexing space, instead of word forms. However, it must be remarked that in contrast to previous approaches, this work used a manually disambiguated test collection (derived from the SemCor (Miller et al., 1993) semantic concordance). Anyway, they measured the sensitivity of retrieval performance to disambiguation errors when indexing documents. For their text collection they found that error rates below 30% still produce better results than standard word indexing, and that from 30% to 60% error rates, it does not behave worse than the standard system indexing with words. They also concluded that if queries are not disambiguated, indexing by synsets performs (at best) only as good as standard word indexing. Nevertheless, it should be mentioned that the topics (queries) had an average of 22 words and were obtained from a summary of each document in the collection. In our opinion, if the topics were not so specific, the benefit of query disambiguation would be questionable.

Finally, an industrial example of sense based retrieval is the theme-based retrieval offered by the "Oracle Intermedia Text Package" (Mahesh et al., 1999). This package offers theme-based retrieval for Document Object Like data by means of the *ABOUT* operator. That is to say, it enables the querying for documents that are about certain themes or concepts. Themes are extracted from documents and queries by parsing them using an extensive lexicon together with a knowledge base of concepts and relations. High precision is achieved by a disambiguation and ranking technique called "theme

proving" whereby a knowledge base relation is verified in the lexical and semantic context of the text in a document. Two themes prove each other if they
are closely connected in the knowledge base either hierarchically or through
cross-references. This eliminates many bad hits arising from word sense am-
biguities.

In general, the conclusion of these works is that WordNet indexing can
improve performance whenever the disambiguation accuracy rate is high (in
some cases not less than 90%). These conclusions are probably not extensible
to the IRS proposed in this chapter because they were indexing free text and
this IRS is constrained to the keywords section of metadata. But from the
results of section 4.3 it can be concluded that the disambiguation accuracy in
our IRS is also quite high (more than 70%). Despite not reaching an accuracy
of 90% for some branches of the thesaurus, it must be taken in mind that
we are disambiguating the own keywords. As opposed to free text retrieval,
we are not going to extract concepts from words that are not essential to the
document meaning. Additionally, most of the works stress that sense based
retrieval seems more appropriate for short query statements. And this is the
case aimed by this information retrieval system: provide access to metadata
catalogs for the general public.

4.4.2 Introduction to the vector-space retrieval model

An information retrieval model can be defined as the specification for the
representation of documents, queries, and the comparison algorithm to retrieve the relevant documents. A formal characterization of such a concept is presented in (Baeza-Yates and Ribeiro-Neto, 1999) as a quadruple
$(D, Q, F, Sim(d_j, q))$ where: D is a set composed of the representation for the
documents in the collection (a collection of metadata records in this case); Q
is a set composed of the representations for the user information needs, called
queries; F is a framework to model document representations, queries and
their relations; and $Sim(d_j, q)$ is a ranking function which associates a real
number with a pair (d_j, q), where $q \in Q$ and $d_j \in D$. Such ranking enables
the ordering of metadata records with regard to the query q.

The vector-space retrieval model (Salton and Lesk, 1968; Salton, 1971)
proposes a framework in which partial matching is possible and it is characterized by the use of a weight vector representing the importance of each index
term with regard to a metadata record (document). Hence, the framework F,
which represents the collection of records and the user queries, consists of
an M-dimensional vector space, where each dimension corresponds with each
distinct index term in the glossary (denoted as T and being M the size of
the glossary). Following expressions show the vector representations of a document d_j and a query q:

$$d_j = ((t_1, w_{1,j}), (t_2, w_{2,j}), \dots, (t_M, w_{M,j})); \quad q = ((t_1, w_{1,q}), (t_2, w_{2,q}), \dots, (t_M, w_{M,q}))$$
$$(4.7)$$

where $t_1, t_2, \dots t_M \in T$ are the M synsets belonging to the glossary; $w_{1,j}, w_{2,j}$,
$\dots w_{M,j}$ represent the weights given to the index terms with respect to d_j;

and $w_{1,q}, w_{2,q}, ...w_{M,q}$ are the weights given to the index terms with respect to q. Finally, this model provides a function to compute the degree of similarity between each metadata record and a user query q, enabling the ranking of records with respect to q. The following equation shows the exact formula to compute the similarity value (denoted as $Sim(d_j, q)$) which is based on the cosine of the angle formed by the vector representing the metadata record and the vector of the user query (Salton and McGill, 1983).

$$Sim(d_j, q) = \frac{\overrightarrow{d_j} \cdot \overrightarrow{q}}{|\overrightarrow{d_j}| \times |\overrightarrow{q}|} = \frac{\sum_{k=1}^{M} w_{k,j} \times w_{k,q}}{\sqrt{\sum_{k=1}^{M} w_{k,j}^2} \times \sqrt{\sum_{k=1}^{M} w_{k,q}^2}} \qquad (4.8)$$

When the vectors d_j and q are equal, they form an angle of 0° and the cosine is 1. On the contrary, an angle of 90° means that the vectors do not coincide in any term and the cosine is 0. The rest of possibilities will indicate a partial matching between the vector representing the metadata record and the vector representing the query: the closer the vectors are in the vector space, the more similar they are. Finally, the aim of the denominator in the previous function is to normalize the result by means of the product of the vector norms. The first norm penalizes metadata records with many index terms. On the contrary, the second norm does not affect the ranking. Next subsections explain the process to obtain the index terms of metadata records and queries.

4.4.3 The indexing of metadata records

Before applying a retrieval algorithm, documents (metadata records) in the collection must be summarized into a set of representative keywords called index terms. In this context of metadata catalogs, metadata records are precisely a summary of media documents (image, text or whatever). Furthermore, the advantage in this context is that metadata creators introduce explicitly the concepts within the keywords section. Nevertheless, the retrieval model of a metadata catalog cannot be based uniquely on a simple matching between a query word and the words contained in keywords section. On one hand, different metadata creators may not share the same criteria to select a harmonized (homogenous) set of keywords. And on the other hand, this simple matching would be comparable with a classic Boolean information retrieval model, where query terms are compared with keywords contained in records to decide whether the record is relevant or not. The disadvantage of this model is that it does not provide any ranking for the relevance of obtained results.

As mentioned in the introduction, one way to increment the descriptive potential of the keywords section is to select terms belonging to formalized controlled lists of terms or thesauri. In this way, more sophisticated methods to resolve terminological queries could be applied. However, there is not a universal thesaurus to classify every type of resource and metadata creators make use of different thesauri or controlled lists depending on the application domain. Therefore, the set of keywords, although using thesauri and controlled lists, are still quite heterogeneous. For example, in the context of geographic information, catalogs may include geographic information about topography, cadastre or communications. Hence, we have proposed the semantic disambiguation of thesaurus terms to avoid this heterogeneity. The main objective of this semantic disambiguation method is to relate the different thesauri to an upper-level ontology like WordNet. Table 4.8 shows the final score of synsets for the branch *accident*, which was displayed in figure 4.6 of section 4.3.2 [2]. The synset with the highest score for each term is elected as the disambiguated synset.

Table 4.8. Disambiguation of a thesaurus branch

Term	Subterm	Synset path	score	lia
accident				
		event→happening→trouble→misfortune→mishap→accident	3,143	0,551
		event→happening→accident	2,560	0,449
accident→accident source				
	accident			
		event→happening→trouble→misfortune→mishap→accident	2,304	0,552
		event→happening→accident	1,873	0,448
	source			
		entity→object→artifact→creation→product→work→publication →reference	0,713	0,231
		entity→object→location→point→beginning	0,705	0,228
		entity→object→artifact→facility→source	0,685	0,221
		entity→life_form→person→communicator→informant	0,397	0,128
		entity→life_form→person→creator→maker→generator	0,397	0,128
		psychological_feature→cognition→content→idea→inspiration →source	0,186	0,060
		abstraction→relation→social_relation→communication →written_communication→writing→document→source	0,009	0,003
accident→accident source→oil slick				
		entity→object→film→oil_slick	0,214	1,000
		...		

Therefore, once a new metadata record has been completed, it is possible to obtain the collection of synsets corresponding to the thesaurus terms. Besides, as the metadata creator probably selected terms from different thesauri, there may be repetition of synsets in the obtained collection. Hence, given the keywords section of a metadata record, it is possible to extract a collection of synsets, which are indeed the index terms and may be characterized by a weight proportional to the number of occurrences and the liability of the disambiguated synset.

As concerns the vector model, one of the best weighting schemes for index terms (the synsets) is the one proposed in (Salton and McGill, 1983), which tries to balance the effect of intra-clustering similarity (features that better describe the documents in a cluster/subset of the collection) and inter-clustering dissimilarity (features which better distinguish a subset/cluster of documents

---

[2] For the sake of clarity, not all the terms and their corresponding synsets have been displayed.

from the remaining documents in the collection) of documents (see equation 4.9). Assuming this weighting scheme, the first step to calculate the weight of a synset is to obtain the frequency of a synset t_i in a metadata record d_j. For a classical information retrieval system, this frequency (denoted as $freq_{i,j}$) would be simply the number of occurrences of an index term. But in this case, we cannot obviate that the disambiguation of thesaurus terms is heuristic and we wanted to consider the score obtained for each synset in the disambiguation process. Therefore, given a thesaurus term s, we have estimated the liability of the elected synset t_i with respect to the other non-elected synsets which were initially associated with the term s. This liability value, denoted as $lia_{s,i}$, is computed as the division between the score of the elected synset and the sum of the scores of all the possible synsets associated with a thesaurus term. Column lia in table 4.8 shows an example of such percentage. $freq_{i,j}$ is finally computed as the sum of the liability of each synset t_i that is indirectly referenced by the terms included in a metadata record d_j. Secondly, it is necessary to obtain the normalized frequency $f_{i,j}$, which is computed as the division between $freq_{i,j}$ and the maximum frequency (computed over all synsets t_l referenced by d_j). Next step is the calculation of the inverse frequency idf_i of a synset t_i, i.e. the logarithm of the division between the size of the collection (denoted as N) and the number of records referencing this synset (denoted as n_i). The point here is that if a synset is referenced in many metadata records, it is not very useful to discriminate them. Finally, the total weight $w_{i,j}$ is computed as the product between $f_{i,j}$ and idf_i.

$$freq_{i,j} = \sum_{s \in d_j} lia_{s,i}; \; f_{i,j} = \frac{freq_{i,j}}{max_{t_l}(freq_{l,j})}; \; idf_i = \log N/n_i; \; w_{i,j} = f_{i,j} \times idf_i; \; (4.9)$$

Additionally, section 4.4.5 (testing of the retrieval method) proposes a variant of the indexing to augment the number of index terms for each metadata record.

4.4.4 The indexing of queries

Regarding the queries formulated by users, it is also necessary to find index terms characterizing these queries. Indeed, the query performed by the user specifies, although vaguely, the set of metadata records that he/she wants to discover. As well as metadata records have been summarized into a collection of synsets, queries must be also synthesized into a set of WordNet synsets. That is to say, in parallel to the indexing of metadata records, every word belonging to the query must be searched into WordNet and then, their possible senses, in the form of synsets, should be processed to obtain a representative collection of synsets. The first question here was whether we should also try the disambiguation of queries or not. By disambiguation of queries it is meant the election of the synset that better represents each query word among its possible synsets found in WordNet. In the context of our experiments it was assumed that the queries contained only a few words and not necessarily connected (i.e. with no synsets in common). Therefore the final decision was the non-disambiguation of queries. Besides, some works like (Voorhees, 1993b) showed that trying to disambiguate the query in addition to the corpus made the results worse, especially in cases where the query was very short. Additionally, it must be mentioned that the use of synsets provides an implicit expansion of query words because each synset represents a set of synonyms (the word typed by the user and all its possible synonyms). (Voorhees, 1993a) essayed different strategies for query expansion also using WordNet synsets and it was concluded that they provided little benefit, at least in the environment (general text retrieval for TREC conference (Voorhees, 2002)) where the experiments were performed.

Finally, regarding query weights, a variant from the weighting scheme in (Salton and McGill, 1983) is applied to compute the weight of every synset with respect to the query q (see eq. 4.10). This variant, suggested in (Salton and Buckley, 1988), gives a minimum weight of 0.5 to the normalized frequency. In this case, $freq_{i,q}$ is computed as the number of indirect references to the synset t_i.

$$w_{i,q} = (0.5 + 0.5 \times (freq_{i,q}/max_{t_l}(freq_{l,q}))) \times idf_i \qquad (4.10)$$

4.4.5 Testing the retrieval model

The information retrieval process

The implementation of this retrieval process is not uniquely a process launched whenever the user performs a query against the catalog. Quite the opposite, this retrieval models, as well as other classic retrieval models, implies a previous work with metadata records contained in the catalog. In fact, the retrieval model involves the following ordered phases (also depicted in figure 4.12):

1. Firstly, the semantic disambiguation of thesauri against the WordNet ontology.
2. Secondly, the creation of metadata records that include terms from disambiguated thesauri in keywords section.
3. Thirdly, the pre-calculation of weight vectors representing the metadata records contained in the catalog.
4. And finally, the computation of the similarity between the query vector and each metadata record whenever the user performs a query.

From the previously mentioned phases, the unique phase that is performed in real time is the fourth one, also in charge of presenting the results to the user. The third phase (metadata indexing) requires a high computational time cost but once it has been finished, it is not repeated unless the content of the catalog is modified (a task not very frequent for stable catalogs).

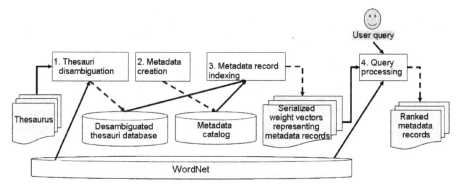

Fig. 4.12. The information retrieval process

Metadata corpus

The formal measures used to quantify retrieval effectiveness of IR systems are based on evaluation experiments conducted under controlled conditions. This requires a testbed comprising a fixed number of documents, a standard set of queries, and relevant and irrelevant documents in the testbed for each query. This is the case of TREC (Voorhees, 2002), an annual conference for academic and industrial text retrieval systems conducted by the National Institute of Standards and Technology, which provides a 2 GB document collection with about half a million documents. However, we could not find such a controlled testbed in the context of metadata catalogs. Therefore, it was necessary the construction of our own testbed.

As an initial metadata corpus, the contents of the Geoscience Data Catalog[3] at the U.S. Geological Survey (USGS) were downloaded. The USGS is the science agency for the U.S. Department of the Interior that provides information about Earth, its natural and living resources, natural hazards, and the environment. And despite being a national agency, it is also sought out by thousands of partners and customers around the world for its natural science expertise and its vast earth and biological data holdings. At the moment of download (March 2003), this catalog contained around 1,000 metadata records describing geospatial data. The metadata records are compliant with the CSDGM standard (FGDC, 1998). Besides, this standard includes a keywords section where the metadata creator can specify different values and the thesauri to which they belong. One of the reasons to select this catalog was our experience with projects dealing with spatial data infrastructures (Bañares et al., 2001; Bernabé et al., 2001, 2002; Gould et al., 2002; Nogueras et al., 2001a; Béjar et al., 2004). However, the results of this work are extensible to any type of digital library using metadata schemas that contains a keyword section or subject attribute. Another important reason to select

---

[3] http://geo-nsdi.er.usgs.gov/

this Geoscience Data Catalog was that it provides a search engine, which is based on ISearch software (Nassar, 1997). ISearch is the search component of ISite, an open source package for indexing and searching documents that implements the Z39.50 information and retrieval protocol (ANSI, 1995). It supports full text and field based searching using the same ranking algorithm as the SMART retrieval system (Salton, 1971), which is precisely the origin of the vector-space retrieval model. This search engine enabled at least the comparison of retrieval effectiveness in terms of qualitative statements and the number of metadata records retrieved.

Once the metadata records were imported in our metadata database, it was found that only 753 of the imported records contained thematic keywords. For our experiments, we were only interested in thematic keywords because WordNet is not specialized on place, temporal or stratum keywords. Furthermore, only 340 of these records contained keywords (an average of 3.673 keywords per record) belonging to formalized thesauri: "National Geologic Map Database Catalog themes, augmented" (NGMDB)[4] with 72 terms appearing 1105 times in the collection; and "Gateway to the Earth" (GTE)[5] with 648 terms appearing only 144 times in the collection. Thus, given that uniquely these thesauri were suitable for the disambiguation, our IRS could use only a small part of the downloaded collection. However, it was noticed that there were 656 records with an average of 7.87 terms belonging to unspecified thesauri, whose name was identified in metadata records as "General" or "none". Therefore, we tried to transform these keywords from unspecified thesauri into terms belonging to GEMET, NGMDB and GTE. In particular, we selected GEMET because, as explained in the testing of the disambiguation method, it is a quite comprehensive thesaurus for geographic information that consists of 5,542 terms organized in 109 branches and translated into 19 languages. In this transformation, we also solved some small morphological differences between the included terms and the terms of the disambiguated thesauri, e.g. difference between singular and plural versions. Table 4.9 displays a summary of the initial and final status of thesaurus terms in each metadata record before and after the aforementioned transformation process: column nt shows the number of terms of each thesaurus, column nr shows the number of metadata records that include these terms; and column avg shows the average number of those terms included by the nr records. The subindex i or f in column headers indicates whether it refers to the initial status or the final status respectively. Thanks to this modification of metadata records, the final collection contained 711 records with an average of 5.594 theme keywords belonging to the three disambiguated thesauri in consideration.

In order to obtain performance measures, a series of topics (queries) and their relevance to metadata records were also necessary. For that purpose, the metadata corpus was enhanced by assigning manually the relevance with

---

[4] http://ngmdb.usgs.gov/

[5] http://alexandria.sdc.ucsb.edu/~ lhill/usgs_terms/usgs/html9/

Table 4.9. Summary of thesauri and theme keywords in the metadata corpus

Thesaurus	nt_i	nr_i	avg_i	nt_f	nr_f	avg_f
AGI Glossary of Geology	17	3	5.67	17	3	5.67
Flouride Environmental Pollution	6	1	6.00	6	1	6.00
GEMET			0.00	1473	520	2.83
General	447	26	17.19	317	26	12.19
GTE	144	25	5.76	1085	407	2.67
NGMDB	1105	329	3.36	1420	598	2.37
None	4716	630	7.49	2805	608	4.61
PrincipalInvestigators	30	15	2.00	30	15	2.00
TOTAL	6465	753	8.59	7153	753	9.50

respect to a series of topics. This way, it would be possible to evaluate the precision and recall of different retrieval systems. The precision is a measure of the ability of a system to present only relevant documents and the recall is a measure of the ability of a system to present all relevant items. The formulas of these measures are the following:

$$precision = \frac{number\ of\ relevant\ hits}{number\ of\ hits} \tag{4.11}$$

$$recall = \frac{number\ of\ relevant\ hits}{number\ of\ relevant\ documents\ in\ collection} \tag{4.12}$$

The topics selected were based on the keywords with highest frequency in the collection. Figure 4.13 displays the 10 topics selected, the thesauri to which they belong and their "narrower term/broader term" relationships. Then, the metadata records were hand-tagged applying two basic rules: "if a specific term a is found in a record m, the record m will be relevant with respect to the broader terms of term a"; and "if a generic term a is found in a record m, the record m will not be relevant with respect to the narrower terms of term a".

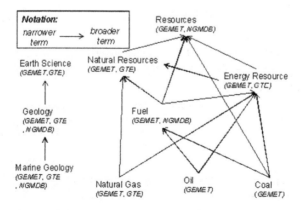

Fig. 4.13. The map of topics

Finally, we wanted to compare the effectiveness of our IRS with respect to a typical word-based retrieval system. But instead of using ISearch for this text retrieval system, the "Oracle Intermedia Text package" (Scherer and Brennan, 2001) was used. Oracle enables the creation of text indexes on text columns that may contain a wide range of Document Object Like data, including XML documents. And by means of the *CONTAINS* operator it is also possible to perform word queries on these columns (including tag based queries for XML documents) and obtain a relevance score. The cause for the replacement of ISearch by Oracle was the disparity in the remote and local data contents. On one hand, the online USGS Catalog updates its contents periodically. And on the other hand, we had modified locally the theme keywords to increment the use of disambiguated thesauri. Anyway, the ranking algorithms of ISearch and Oracle are very similar. To obtain the relevance score, both systems use an inverse frequency algorithm based on the vector-space model formulas. In fact, before the transformation of keywords, a series of tests were performed against online ISearch and Oracle (containing same records in XML) and equivalent results were obtained. For the comparison of experiments with Oracle, all the keywords (from disambiguated thesauri) of each record were comprised into a large text column value (CLOB). And then, an Oracle text index was created on this text column.

The experiments

After analyzing the synsets that are referenced indirectly by the metadata records of our catalog, we obtained that 201 synsets were referenced by 707 metadata records, each record referencing an average of 4.065 synsets and 20 synsets at maximum. And the minimum, maximum and average values for n_i were 1, 15.228 and 347.

As a first search example, the query *geology erosion* was performed obtaining the results presented in table 4.10 (ordered by similarity). And as an example, the computation of the similarity between the query and the first result is shown. The query contains two words that are associated with 5 WordNet synsets, whose weights are displayed in table 4.11. The weights of synsets referenced by the first 3 hits are presented in table 4.12 and the similarity for the first metadata record is given by

$$Sim(d_j, q) = \frac{5.46 \times 3.27}{\sqrt{3.27^2 + 5.46^2 + 5.86^2} \times \sqrt{0.89^2 + 0.71^2 + 0^2 + 5.46^2 + 0^2}} = 0.37$$

$$(4.13)$$

One effect that can be observed from the results obtained with the query *geology erosion* is the influence of the inverse frequency and the number of keywords used in each metadata record. For instance, that is the reason to explain the relevance of the metadata record in 3rd position. Although, it references three synsets (*10413485, 4655198* and *6691504*) that match with the

Table 4.10. Hits for the query *geology erosion*

Order	Title	Sim
1	Beach profile data for Maui, Hawaii	0.375
2	Beach profile data for Oahu, Hawaii	0.375
3	Possible Costs Associated with Investigating and Mitigating Some Geologic Hazards in Rural Parts of San Mateo County, California	0.318
...		

Table 4.11. Computation of synset weights for the query *geology erosion*

Word	Synset	$freq_{i,q}$	$f_{i,q}$	n_i	$w_{i,q} = (0.5 + 0.5f_i, q) \times idf_i$
Geology	4655198 (a science that deals with the history of the earth as recorded in rocks)	1	1/1	288	1 x ln(707/288)=0.89
	6691504 (geological features of the earth)	1	1/1	347	1 x ln(707/347)=0.71
Erosion	9691024 (the mechanical process of wearing or grinding something down)	1	1/1	0	0
	10413485 (condition in which the earth's surface is worn away by the action of water and wind)	1	1/1	3	1 x ln(707/3)=5.46
	9691547 (erosion by chemical action)	1	1/1	0	0

Table 4.12. Computation of synset weights for the first 3 hits (N = 707)

Order	Thesaurus, Keyword	synset	lia_i	$freq_{i,j}$	$f_{i,j}$	n_i	idf_i	$w_{i,j}$
1, 2	GEMET, Coastal Erosion	10413485	0.6	0.6	0.6/1	3	5.46	3.27
		6801422	1	1	1/1	3	5.46	5.46
	GEMET, Beach	6739108	1	1	1/1	2	5.87	5.86
3	GEMET,landslide GTE,landslides	5512262	0.36 0.53	0.89	0.89/3	0.89/3	4.48	1.627
	GEMET,earthquake GTE,earthquakes NGMDB,earthquakes	5526375	1 1 1	3	3/3	10	4.26	4.259
	GEMET,coastal erosion GTE,erosion	10413485	0.6 0.46	1.06	1.06/3	3	5.46	1.941
	GEMET,coastal erosion	6801422	1	1	1/3	3	5.46	1.821
	GEMET,cost	4008333	0.33	0.33	0.33/3	1	6.56	0.7419
	GEMET,slope	6724958	1	1	1/3	1	6.56	1,982
	GTE,structural geology GTE,geology	4655198	0.66 0.6	1.26	1.26/3	288	0.9	0.385
	GTE,structural geology	4655855	0.21	0.21	0.21/3	16	3.79	0.291
	GTE,fracture(geologic)	6691504	1	1	1/3	347	0.71	0.236
		6735707	0.39	0.39	0.39/3	14	3.92	0.513
	GTE,liquefaction	9738666	1	1	1/3	1	6.56	2.187
	GTE,maps and atlases	2965788	0.54	0.54	0.54/3	18	3.67	0.687
		4843693	0.39	0.39	0.39/3	18	3.67	0.482

query synsets, its similarity to the query is lower than similarity of the first two hits (having only one match with the query synsets). On one hand, two of the synset matches (*4655198* and *6691504*) correspond to the synsets associated with *geology*, whose inverse frequency is very low. These synsets are very frequent in the collection and the weighting scheme used tries to balance this effect: "the fewer a term occurs in, the more important it must be". Sometimes this is not satisfactory, but more often it is useful. And on the other hand, the record in third position references a total number of 13 synsets, while the first two hits reference only 3. As the number of referenced synsets grows, the norm of the vector representing the record will increase, increasing as well the denominator in the similarity formula. This denominator favors metadata records with fewer keywords. Although some times this means that such metadata records are better focused on a subject, other times is simply due to a worse quality in metadata cataloguing. It was tested the possibility of obviating the denominator (always equals to one). But this variation was rejected because the results were not satisfactory: there was almost no graduation (a great deal of hits shared the same similarity value) for the similarity in simple queries as the previous one. Besides, as the number of query terms and synset matches increases, the norm of the vectors representing the records is not so influent.

Then, we wanted to test one of the obvious advantages of our information retrieval system in comparison with the ISearch software. It is that the queries can contain words that have not been necessarily included in metadata keywords, e.g. synonyms of these keywords that match with the same WordNet synsets. For instance, we performed two queries with two synonyms, *fuel* (or *fuels*) and *combustible*, which correspond to the same WordNet synset (*10669661, a substance that can be burned to ...*). Our IRS always returned 138 hits but Oracle only returned records (138 hits with same score) for the query *fuels*, which was the word included in the keyword section. Basically, the hits returned by our IRS were graduated by the number of synsets indirectly referenced: 2 synsets for the first 18 hits, 3 synsets for the following 35 hits and so on. Table 4.13 shows the first distinct similarity values for the hits of the query *fuel* with our IRS.

Table 4.13. Hits for the query *fuel*

Order	Title	Sim
1	Coal Bearing Regions and Structural Sedimentary Basins of China and Adjacent Seas: Major coal mine locations	0.748
...	...	...
19	jbcat.shp (net coal thickness in Deadman coal zone, Jim Bridger area)	0.698
...	...	...
54	Oil and Gas Field Centerpoints of Australia and New Zealand (fld_3anz)	0.682
...	...	...
58	shrbeds (Tertiaryaged coal beds in the Sheridan coalfield)	0.669
...	...	...

After these initial experiments, we decided to augment the number of synsets representing the metadata records. For this expansion, we included the disambiguated synsets that were associated to the broader terms of the terms included in the *keywords* section. For instance, the broader term of *coal* in GEMET thesaurus is *fossil fuel*, and thus metadata records with term *coal* would be indexed with the disambiguated synset of *coal* (*10628288, carbonized vegetable matter deposited in ...*) as well as with the disambiguated synset of *fossil fuel* (*10527530, fuel consisting of the remains of organisms preserved in rocks ...*). The idea was that if a user asks for resources about *fossil fuel*,

Table 4.14. Hits for the query *fossil fuel*

Order	Title	Sim
1	dan_pts (Public data points in Danforth Hills coal field)	0.698
2	csb_bnd* (The outcrop and area underlain by the John Henry Member of the Straight Cliffs Fm. in the Kaiparowits Plateau study area, southern Utah)	0.698
3	csb_strc (The structure contours of the Calico sequence boundary in the Kaiparowits Plateau, southern Utah)	0.698
4	kai_adit (Coal mine adits within the Kaiparowits Plateau study area, southern Utah)	0.698
. . .	. . .	. . .
28	kai_strc (Geologic structural features within the Kaiparowits Plateau study area, southern Utah)	0.535
. . .	. . .	. . .

he might be interested in different types of fossil fuels (e.g. *coal, natural gas* or *petroleum*). Of course, the weight of the synset for broader term must be lower than the weight for the real term included in the metadata record. In particular, the liability of the synsets which are associated with broader terms is divided by 2. With this expansion we obtained that 272 synsets were referenced by 709 metadata records, each record referencing an average of 5.988 synsets and 29 synsets at maximum. And the minimum, maximum and average values for n_i were 1, 16.577 and 362. For instance, thanks to this modification, our IRS returned 121 hits for the query *fossil fuel*, one hit more than the query *coal*. Meanwhile, Oracle returns no hits for query *fossil fuel*. This is due to the fact that Oracle *CONTAINS* operator only performs simple word matching, and only the word *fuels* (the plural version of *fuel*) is included in metadata records. Some results obtained with our IRS for the query *fossil fuel* are shown in table 4.14. Table 4.15 shows the weights of synsets for the first 27 hits (having same keywords) of the query *fossil fuel*. There, it can be observed that the liability of synset *10527530 (fossil fuel)* has been divided by 2. On the other hand, the query *fossil fuel* references uniquely the synset *10527530* with the weight 1.76 $((0.5 + ((0.5 \times 1)/1) \times ln(709/121))$. Therefore, the similarity for the first 27 hits can be computed as follows:

$$Sim(d_j, q) = \frac{1.73 \times 1.76}{\sqrt{1.77^2 + 1.73^2} \times \sqrt{1.76^2}} = 0.698 \qquad (4.14)$$

Table 4.15. Weights for the first 27 hits obtained with the query *fossil fuel* (N=709)

Order	Thesaurus, Keyword	synset	lia_i	$freq_{i,j}$	$f_{i,j}$	n_i	idf_i	$w_{i,j}$
1-27	GEMET, Coal	10628288 (coal)	0.5	0.5	0.5/0.5	120	1.77	1.77
		10527530 (fossil fuel)	0.98/2	0.49	0.49/0.5	121	1.77	1.73

Finally, we compared the performance of the basic indexing of our IRS, the extended indexing of our IRS and the Oracle text retrieval. Figure 4.14 displays the average precision-recall curves obtained with the aforementioned topics and for the different types of retrieval systems. A precision-recall curve interpolates precision numbers against percentage recall values. For instance,

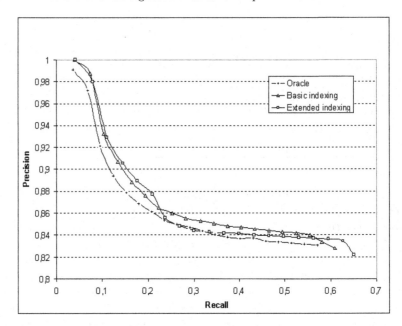

Fig. 4.14. Average precision-recall curves

a percentage recall of 50% is the position in the hitlist at which 50% of the relevant documents in the collection have been retrieved. As mentioned in (van Rijsbergen, 1979), it is an experimental fact that precision-recall curves are monotonically decreasing: the increase of recall usually implies the inclusion of some bad hits within the hitlist. To measure the overall performance of a retrieval system, the set of curves, one for each topic query, must be combined in some way to produce an average curve. In particular, the average curves of figure 4.14 were obtained by means of the micro-evaluation algorithm presented in (van Rijsbergen, 1979).

Basically, it can be concluded that the precision obtained in the tests is similar for the both three cases: the precision fall of the basic indexing with respect to Oracle results is 0.29%; and in the case of extended indexing it decreases only 1.06%. On the other hand, the main advantage of the IRS proposed in this chapter is that the recall measures are improved: an increase of 6.60% in the case of basic indexing with respect to Oracle; and an increase of 13.94% in the case of extended indexing.

4.5 Conclusions and future work

This chapter has presented in first place an unsupervised technique for the disambiguation of thesaurus terms in light of their surrounding terms and with reference to an external upper level ontology, WordNet in this particular case. This method takes advantage of the thesaurus hierarchical structure (broader and narrower terms), which is used as the word context for a voting algorithm to find the closest sense.

The main disadvantage of this disambiguation method is that it may not be adequate for the semantic disambiguation of very specific domain ontologies because WordNet lacks for domain-specific terminology. Nevertheless, the intention of this work is to approximate as much as possible the terms used in metadata records and the concepts extracted from "general-purpose" queries. Furthermore, this disambiguation algorithm provides the basis for a wide range of interesting applications.

This chapter has presented one of these applications in the adaptation of a classic information retrieval model in the context of a digital library, understood as a catalog holding metadata records. In particular, the applicability of the vector-space model has been explored. In more heterogeneous contexts, other retrieval models, such as probabilistic or neural-net based models, would work probably better. However, in this context of metadata catalogs, the own metadata records are the summary of the desired resource and a simple model may provide satisfactory results.

Regarding the indexing of metadata records, it has been assumed that the metadata schema includes a keyword section or subject element, something quite usual in most metadata schemas. Besides, the indexing technique is based on the inclusion in this section of terms selected from disambiguated thesauri. The index terms are precisely the synsets associated with the selected thesaurus term during the disambiguation process of the thesaurus. Furthermore, this basic indexing of metadata records was modified to augment the number of index terms. Apart from collecting the synsets associated with a thesaurus term, the indexing method also included the synsets associated with the broader terms in the thesaurus hierarchy. These synsets coming from broader terms were assigned a lower weight. This modification was based on the assumption that metadata records represented by these synsets (from broader terms) are still semantically close to queries including the broader concept. This expansion could have been also continued with the synsets associated with other related terms. However, works like (Clark et al., 2000) suggest not considering concepts at distance two or more from an initial concept.

The viability of the retrieval model has been tested with a collection of metadata records describing geographic resources and the results have been compared with a typical text retrieval system (based on word matching). These first experiments have shown that the precision obtained is comparable with a typical text retrieval system. The main advantage of the information retrieval system presented in this chapter is the increment in the number of relevant documents returned, i.e. the improvement of recall measures. Anyway, it is necessary to test the method with a bigger corpus of metadata records and better classified with additional disambiguated thesauri.

An improvement in the computation of the weight of each index term would be to consider the importance of a thesaurus, to which the terms in the keyword section belong. A term selected from a specific thesaurus like "GEMET" may be more relevant than a term belonging to a thesaurus that compiles only a hundred of categories. Another improvement in the method could be a better representation of the user query. Apart from the synsets related with words contained in the query phrase, the ancestors of these synsets in WordNet hierarchy could be also considered. In this way, the information retrieval system could return, at least, metadata records referencing synsets in the ancestor hierarchy of the query synsets. Furthermore, the words contained in the definition of the WordNet synsets could be used to expand the query formulated by the user. Additionally, it must be mentioned that this retrieval method could be extended by indexing other metadata fields (or elements) like *title*, or *abstract*. Besides, the value of similarity could be integrated into more complex information retrieval systems as another factor to compute the final value for the degree of similarity.

And finally, we must remark that this chapter has only addressed the interconnection of English thesauri. One of the remaining tasks of this disambiguation algorithm is to incorporate the Spanish WordNet. The use of EuroWordNet, which connects WordNets in different languages, is a promising approach to allow cross-language (multilingual) text retrieval as proposed by works like (Gonzalo et al., 1998b).

5

Integrating the concepts within the components of a Spatial Data Infrastructure

5.1 Introduction

This chapter is devoted to demonstrate that the concepts presented in previous chapters have contributed to the improvement of some of the components that integrate a spatial data infrastructure, and that have been deployed within real use cases of spatial data infrastructures.

In order to remember the technical components of a spatial data infrastructure, figure 5.1 shows the architecture of a prototypical spatial data infrastructure that was already presented in chapter 1. There, four main areas of components were identified: a geographic catalog area including services and applications that contribute to publish the descriptions of the geographic resources available; a set of services facilitating the visualization, access and geoprocessing of geographic information; a services catalog publicizing the description of services offered at the SDI node; and a series of client applications making use of the data and services offered by the spatial data infrastructure. And within all these possible components, the work presented along this book has actively contributed to increase the capacity of three of them: the creation of catalog services (both describing resources and services), the development of enhanced metadata editors, and the construction of portals facilitating the access to the services and resources offered by an SDI node.

On one hand, the proposals for collections management support, the availability of crosswalks between metadata standards and the sense based information based retrieval have enabled the development of a versatile catalog services component. As explained in chapter 2 the catalog services have been built over a Metadata Knowledge Base that gives support for collections, facilitates automatic inference of metadata and provides intelligent query answering according to the hierarchical structure of collections. But in addition to this, the Metadata Knowledge Base has also incorporated the use of crosswalks (developed according to the process explained in chapter 3) and the sense based retrieval strategies from chapter 4. The former contribution has facilitated the interoperability between metadata standards, i.e. it has en-

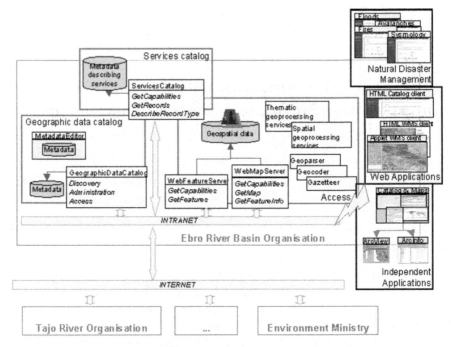

Fig. 5.1. Architecture

abled a multistandard catalog that manages transparently metadata entries with independence of its original standard. And the latter contribution, based on the disambiguation of thesaurus terms, has provided an additional factor to compute the relevance of a metadata record with respect to the user query.

On the other hand, the catalog software has been integrated into a metadata edition tool called *CatMDEdit*. Apart from the basic management of metadata entries in the local repository of the catalog, the integration of this software has facilitated the incorporation of additional functionalities in the metadata edition tool: the unified edition of metadata for collections; the interoperability between different metadata standards; and the integration of a thesaurus module for selecting terms of disambiguated thesauri.

And finally, different customized search interfaces accessing the catalog services have been developed and integrated within the Web Portals of distinct spatial data infrastructures. These search interfaces are usually characterized by providing presentation of metadata according to different standards, and by providing the navigation through collection of related resources.

The rest of the chapter is structured as follows. The next section will present the improved capacities of catalog services. Then, section 5.3 presents the features of an enhanced metadata editor. Section 5.4 presents the construction of the Web portal of a spatial data infrastructure. And finally, this chapters ends with some conclusions and future work.

5.2 The catalog services component

5.2.1 Introduction

Figure 5.2 shows the architecture of the catalog system that has been already presented in chapter 2. But this time, apart from the use of the Metadata Knowledge Base proposed in that chapter, we have also remarked the contributions presented by chapters 3 and 4 to the catalog server (deployment version of the catalog services component) and its client applications.

First of all, it must be mentioned that this enhanced catalog services component supersedes a previous version, which enabled a basic management of metadata entries and whose details (most of them still valid for this present enhanced component) can be found in (Zarazaga et al., 2000b; Nogueras et al., 2001b; Cantán et al., 2000). The implementation of the services of this initial version consisted in the direct access to a relational database (e.g., Oracle or Access accessed via JDBC [1]), which stored the different sections of metadata records in a set of related tables. However, this implementation approach proved to be not flexible and in some cases inefficient. On one hand, it did not allow the support for multiple metadata standards (every standard required its specific relational model). And on the other hand, the response time of catalog searches was relatively high due to the execution of selects in multiple relational tables.

Therefore, the development of the catalog services over a Metadata Knowledge Base has represented a decisive change. As it was explained in chapter 2, the use of a Metadata Knowledge Base has allowed the storage of metadata according to different standards and the management of nested collections. But in addition to this, this Metadata Knowledge Base has also incorporated the use of crosswalks to support metadata interoperability in the catalog (chapter 3) and the concept-based information retrieval strategy (chapter 4). Thus, these additional capacities of the catalog services component have had a direct influence in the improvement of the rest of components that integrate or make use of the catalog services. Firstly, the catalog server offers a standardized interface (*Standard Web Interface* in figure 5.2) for the discovery services according to the OGC Catalog Interface Implementation Specification (Nebert, 2002). And the difference of this interface implementation with respect to other implementations is the ability of supporting query and presentation of results according to different metadata standards. This is possible

---

[1] As well as the rest of software referred in this work, the catalog services component has been developed in Java. JDBC (Java Database Connectivity) provides a standardized application programming interface for the access to relational databases.

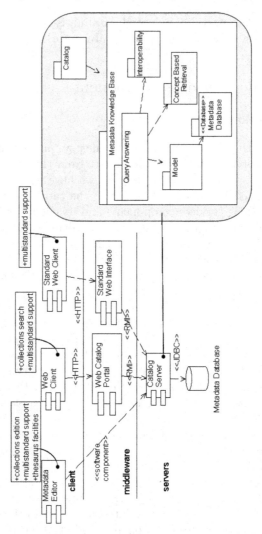

Fig. 5.2. Architecture of the catalog services component

thanks to the integration of crosswalks in the catalog and the flexibility of
the Metadata Knowledge Base to store different types of metadata. Secondly,
the catalog software has been integrated into a metadata edition tool called
CatMDEdit. This tool, presented in section 5.3, enables as special features:
the unified edition of metadata for collections; the interoperability between
different metadata standards; and the integration of a thesaurus module for
selecting terms of disambiguated thesauri. And thirdly, different customized
search interfaces accessing the catalog server have been developed and inte-

grated within the Web Portals of distinct spatial data infrastructures. The distinctive features of these portals, explained in section 5.4, are the presentation of metadata according to different standards and the navigation through collection of related resources.

Next subsections detail how the metadata standard interoperability and the concept based retrieval have been integrated within this catalog services component.

5.2.2 Integrating the interoperability between metadata standards

Chapter 3 presented a process for the construction of crosswalks between metadata different metadata standards. As an example of the construction of crosswalks, several crosswalks were developed to enable the interoperability among different geographic metadata standards such as CSDGM (FGDC, 1998), ISO 19115 (ISO, 2003a) or MIGRA (AENOR, 1998); and also to enable the interoperability with more general standards like Dublin Core (DCMI, 2004; ISO, 2003d; ANSI, 2001). This section details now how these crosswalks have been integrated within the catalog software to enable the desired metadata interoperability.

As it was mentioned in chapter 2, the Metadata Knowledge Base was designed to support the storage of metadata records in conformance with different standards. However, in order to enable users to make queries with independence of the metadata standard, two issues had to be solved. On one hand, the client restrictions on the metadata elements of a specific standard had to be translated/expanded to all the metadata standards that were used by the metadata entries stored in the catalog repository. And on the other hand, the metadata records obtained as a result of the user query had to be converted to the standard specified by the user.

Figure 5.3 shows how the crosswalks have been integrated within the query answering component of the Metadata Knowledge Base. The *CrosswalkBroker* class provides the necessary functionality as follows:

- The *expandQuery* method expands the user query to a new query that takes into account all the metadata standards supported by the catalog. The initial user query can be considered as a query tree whose nodes are logical operators (*and, or*) and where the leaves are restrictions on the elements of a specific standard (e.g., comparison operators like $=, \neq, >, <$ or *like* on the values of an element). This method will transform each leaf of this query tree into a query tree that consists of the disjunction (*or*) of the user restriction transformed to all the possible standards supported by the catalog. That is to say, given the original element name specified by the user, the method will find the corresponding element in the rest of standards (the labor of the *convertElement* method) and will generate an equivalent restriction. This is done thanks to the existence of the associations between elements that were specified during the semantic mapping phase of crosswalks construction.

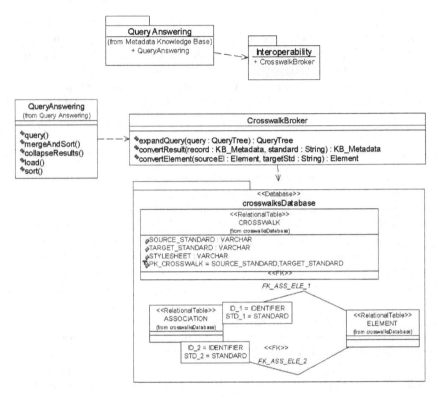

Fig. 5.3. Integration of crosswalks

- On the other hand, the *convertResult* method allows the conversion of the metadata records retrieved as a result of a user query into the desired standard specified by the user. Thus, this method must apply the appropriate stylesheet that converts the internal XML metadata into the desired standard. For that purpose, this method accesses a crosswalk repository that contains the crosswalks that are available and selects the appropriate one.

Different client applications of the catalog make profit of this metadata interoperability. A clear example are the client and server implementations of the Web profile of the OGC Catalog Services Specification. This specification defines the syntax of the messages used for client requests and server responses, which are encoded in XML and transferred using the HTTP protocol. And among these messages, the *search request* message includes an attribute called *preferredRecordSyntax* that indicates the format of the records returned by the catalog server. That is to say, it specifies the metadata standard, and consequently, it indicates the XML Schema associated to that standard that will be used to present the records to the user.

Fig. 5.4. Multi-standard support in the OGC Catalog Client-Server interaction

In order to illustrate this interoperability, figure 5.4 shows a Java applet implementing the client of this catalog services specification. This client offers a simple graphical user interface whose main objective is to validate the correct implementation and operation of catalog services. This interface is structured in three tabbed panels: the *Main* tab to facilitate the construction of requests, the *Request* tab to display the complete request sent to the server, and the *Response* tab to visualize the server responses. There, it can be observed how a user has created a search request in the *Main* tab specifying ISO (meaning ISO 19115) as standard and a restriction that filters all the records containing the word *hydrology* as a keyword (see *"Filter:"* text area). Once the user press the button *Search*, a *search request* is sent to the server and it can visualized in the *Request* tab. In the figure it can be observed that the *preferredRecordSyntax* field contains the value *ISO*. And finally, after performing the *search request*, the user may perform a *present request* to obtain some of the records that verify the restriction included in the search request. The tab *Response* of figure 5.4 displays part of a retrieved metadata record in XML format and in conformance with the ISO 19115 XML-Schema. Further details about the implementation of the OGC Catalog services specification can be found in (Muro-Medrano et al., 2003; Nogueras-Iso et al., 2004c).

Finally, it must be mentioned that the crosswalk broker depicted in figure 5.3 does not need to be so tightly coupled with the catalog implementation. For instance, this crosswalk broker could be used in the construction of a gateway giving access to a network of distributed catalogs independently of their implementation or supported standards. The gateway would use the crosswalk broker to translate the user queries to the standard supported by each catalog in the network. And before presenting the records returned by each catalog, the gateway would also make use of the crosswalk broker to convert each record to the standard/schema desired by the user. In fact, we are describing the way to implement the use-case presented in the introduction section of chapter 3, i.e. the case of a tourism agency that had to merge the information from three different metadata databases (see figure 3.1).

5.2.3 Integrating the concept based information retrieval

Chapter 4 presented a concept-based information retrieval strategy that was based on the disambiguation of thesauri with respect to the WordNet upper-level lexical database. The objective here is to present how this strategy has been incorporated within the Metadata Knowledge Base to provide an additional factor to compute the relevance of a metadata record with respect to the user query.

Figure 5.5 shows the integration of the concept based retrieval within the Query Answering component of the Metadata Knowledge Base. This integration can be summed up as follows:

- On one hand, the *IndexGenerator* class has the responsibility of pre-calculating weight vectors representing the metadata records contained

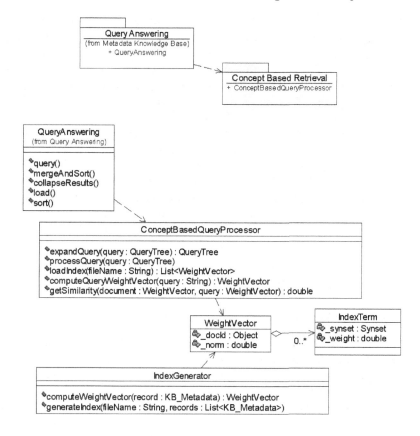

Fig. 5.5. Integration of the concept based information retrieval

in the catalog. The *generateIndex* method receives a series of metadata records, obtains the weight vector of each metadata record (by means of the *computeWeightVector* method), and serializes the index terms of each metadata record to a file. Although this task requires a high computational time cost, it is performed off-line. Once it has been finished, it is not repeated unless the content of the catalog is modified (a task not very frequent for stable catalogs).

- And on the other hand, the *ConceptBasedQueryProcessor* class computes the similarity between the query vector and each metadata record whenever the user performs a query. The *loadIndex* method reconstructs the index, which had been previously computed by the *IndexGenerator* class. The *expandQuery* method expands the initial user query into a new query that takes into account all the concept-based restrictions specified. Similar to the way of working of the method with the same name in the *Crosswalks Broker* class of section 5.2.2, this method will filter those leafs of the query tree specifying restrictions on elements containing keywords (e.g., element *subject* in Dublin Core or *descriptiveKeywords* in ISO 19115) and will transform them into new query trees. These new query trees will consist of the disjunction (*or*) of the initial restriction and a new restriction using a special operator called *conceptLike*, which will denote that the *QueryAnswering* class must make use of the *ConceptBasedQueryProcessor* to process this restriction. The *processQuery* method will process then a restriction using this *conceptLike* special operator. Firstly, this method will use the *computeQueryWeightVector* method to obtain the weight vector of the query. And secondly, it will check the similarity between each metadata record and the user query by means of the *getSimilarity* method. The metadata records ranked with a similarity greater than 0 are attached to the query tree received as parameter and they will be later merged and sorted with the rest of results (as it was explained in section 2.4.4).

5.3 A metadata editor

5.3.1 Introduction

One of the main problems for launching a spatial data infrastructure is to have appropriate and well-defined contents for its catalogs. The creation of metadata is an arduous labor that must be facilitated by the adequate tools.

A revision of metadata edition tools for geographic metadata can be found in appendix C.1. All of them have a series of common characteristics that could be summed up as follows: the edition of metadata records according to a metadata standard (e.g., CSDGM or ISO 19115); the possibility of validating the consistency (correctness) of metadata records (e.g., check the obligation and maximum/minimum number of occurrences of each element); the exchange of metadata records in a standardized format (XML conforming to the DTD or XML-Schema established by the standard); and presentation of metadata records in human readable formats like text or HTML and with different styles (FGDC, FAQ, GeographyNetwork, ESRI). Additionally, other desirable features of metadata edition tools could be: mechanisms to facilitate classification (provide selected vocabulary to facilitate later searching); mechanisms for the automatic generation of metadata; support for the internationalization and coordination of multilingual versions of metadata records; and independence of the tool with respect to the database vendor (or storage device) and the execution platform.

Integrating the catalog as another software component, we have also created a metadata edition tool, which is called *CatMDEdit* (Zarazaga et al., 2000c; Zarazaga-Soria et al., 2003a). It has been implemented in Java and complies with all the characteristics of metadata that we have stated before. But apart from this basic functionality, it facilitates: the joint cataloguing of collections of resources; the interoperability between different metadata

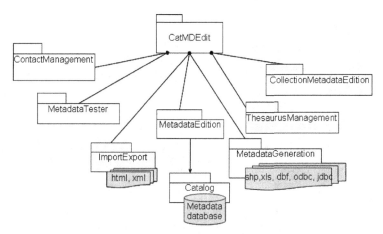

Fig. 5.6. Components of *CatMDEdit*

standards; and the integration of a thesaurus module for selecting terms of disambiguated thesauri. Figure 5.6 shows the different components that form part of the *CatMDEdit* tool. They are the following:

- The *Catalog* component enables the storage of metadata entries in a relational database. This component is the same software that makes use of the Metadata Knowledge Base and that we have already detailed in chapter 2.
- The *MetadataEdition* component, as its own name indicates, enables the edition and visualization of metadata entries. Not only does it supports the edition of metadata records in compliance with geographic metadata standards like CSDGM and ISO 19115, but it also allows the edition more general metadata standards like Dublin Core. Additionally, it facilitates the creation of multilingual versions of metadata records, translating automatically as much content as possible.
- The *MetadataTester* component facilitates the validation of metadata elements of a metadata record inserted by the user until that moment. That is to say, it checks the obligation (mandatory, optional, or conditional) and minimum/maximum occurrences of the metadata elements according to a specific standard and reminds which mandatory elements have not been filled in yet.
- The *ContactManagement* component permits the reuse of contact information (e.g. name, address, telephone ...) of organization and individuals, which must be filled in several metadata elements. Thanks to this component, the contact information about a responsible party is inserted only once and used whenever it is required.
- The *MetadataGeneration* component is able to derive metadata from the data sources by means of interconnection with commercial GIS tools or

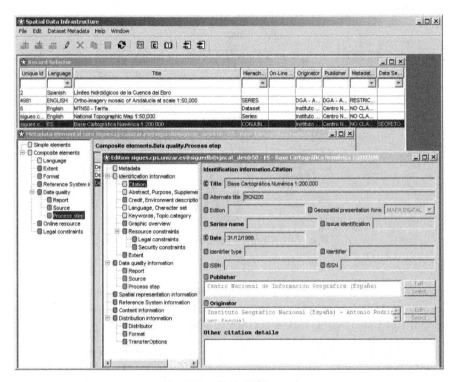

Fig. 5.7. *CatMDEdit* tool

proprietary software. Examples of derived metadata are information about spatial reference systems, number and type of geographic features, extension covered by a record, or information about the entities and attributes of alphanumerical related data.

- The *ImportExport* component enables the exchange of metadata records according to different standards and formats. It is possible to specify a standard different from the original standard of the selected metadata record.

- The *ThesaurusManagement* component enables metadata creators to use thesauri in order to fill in some metadata elements. The use of these controlled keywords facilitates the mapping between a selected vocabulary and a large collection of records. This way, the catalog discovery services may guide the discovery of datasets by using hierarchies of concepts. In addition to this, this thesaurus component implements the disambiguation method presented in chapter 4 and other applications based on this disambiguation.

- And the *CollectionsMetadataEdition* component facilitates the creation of metadata for collections of geographic resources. This component provides the interface to access the functionality offered by the catalog for the management of collections (already presented in chapter 2).

Figure 5.7 displays the graphic user interface of this tool. From all the components of this application, the last three ones integrate some of the concepts introduced by this book and will be detailed in next subsections. As concerns the rest of the functionality of the tool, further details can be found at (Zarazaga et al., 2000c; Zarazaga-Soria et al., 2003a).

5.3.2 Import/Export of metadata

The *ImportExport* component enables the exchange of metadata records in XML format (tagged plain text files) conforming to different standards such as CSDGM, ISO 19115 and Dublin Core. The use of agreed standards facilitates the understanding and interoperation with other applications making use of metadata.

There are two forms in the CatMDEdit tool to insert a new record: adding a new empty record and importing the contents of the new record from an XML file. In both cases, the user can specify the standard according to which the metadata will be stored in the database. In the case of adding a new empty record, there is no problem. But in the case of importing an XML file whose standard is not the same as the standard selected by the user, a translation must be performed. In this second case, a crosswalk must be applied to the source XML file in order to obtain the metadata contents adjusted to the desired standard. Equally, when a user exports a record in XML format and according to a standard different from the original standard of metadata, the necessary croswalks must be applied to obtain the target standard.

Finally, it must be mentioned that this component also facilitates the generation of more readable presentations of metadata records in HTML format, e.g. English and Spanish FAQ, ESRI, and Geography Network style presentations. For instance, figure 5.8 shows the same metadata record displayed according to different types of HTML presentations.

5.3.3 Collection Metadata Edition

The *CollectionMetadataEdition* component enables the edition and visualization of metadata describing collections of datasets that can be considered as a unique entity. At present, it only enables the edition of spatial collections, but new types of collections will be supported in the future. As mentioned in chapter 2, the components of spatial collection collections are spatially distributed and they have usually arisen as a result of the fragmentation of geographic resources into datasets of manageable size and similar scale. For instance, examples of such collections are mosaics of aerial images or national topographic maps which are divided into tiles of equal area at a concrete scale.

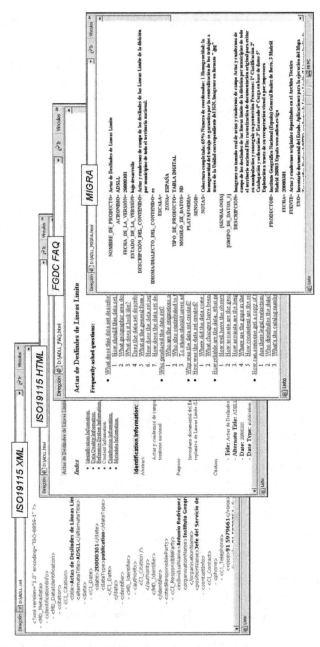

Fig. 5.8. Different styles of presentations

The objective of this tool is to manage jointly the metadata at collection level (shared by the all the components in the collection) and the specific

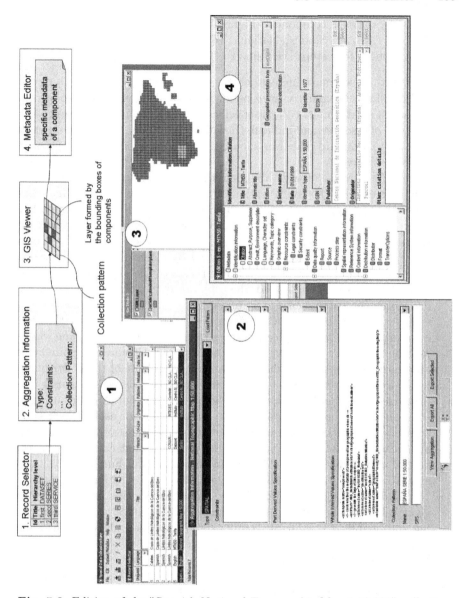

Fig. 5.9. Edition of the "Spanish National Topographic Map 1:50,000" collection

characteristics of each component. Figure 5.9 illustrates the process of editing a collection (e.g., the "Spanish National Topographic Map at 1:50,000 scale"), which consists of four steps according to the present prototype:

1. Firstly, the user must indicate that the metadata record describing the entire collection explicitly corresponds with a collection. That is to say, the metadata creator must edit the specific metadata element that categorizes the type of resource and indicate that it is a collection. For instance, using ISO 19115 standard, the user should fill the element *hierarchy level* with the value *"series"*.

2. Then, the user is allowed to open the *Aggregation Information* window that enables the configuration of the aggregation relation that associates the metadata record describing the collection with all the metadata records describing the components. In this window the user will select the type of aggregation relation that is used in this collection (at present, only the *SpatialRelation* type is allowed). And by selecting one of these pre-established types, most of the characteristics of the aggregation relation type will be directly configured: the *constraints* that the metadata records describing the components must observe (e.g., the *geographicLocation* element must not be null), the *wholeInferredValuesSpecification* that will compute values for the collection record, and the *partDerivedValuesSpecification* that will compute values for the component records. Despite being automatically configured, the window shows the value of these attributes, which are encoded in XSL, for verification purposes. Finally, the user may also select the spatial pattern that explains the spatial distribution of components in the collection and that is particular for each collection instance. This spatial pattern can be expressed, for instance, by means of an ArcView SHAPE file containing the polygons that correspond to the spatial extent of each possible unit in the collection. Furthermore, these spatial patterns are typically reused for different collections. Frequently, National Geographic Institutes define spatial distribution patterns for core geographic data at different scales, thus providing an established numbering and bounding box for the components (also called tiles). For instance, in figure 5.9 the grid defining the spatial extent of tiles at 1:50,000 scale was selected.

3. The *CollectionMetadataEdition* component includes a visualization tool that enables the supervision of the status of cataloguing. The tool generates a GML layer (Cox et al., 2003) whose geographic features correspond with the records describing the components of the collection as follows: the geographic location of each feature corresponds with the bounding box that defines the value of the *geographicLocation* element of a component record; and the rest of feature attributes store the necessary identifiers defining the link to the metadata record of each component. Thanks to this, this tool can provide an approximate view of what has been already catalogued.

4. And last, through the visualization tool, it is possible to select a component and open the window that enables the edition of the specific metadata of this component.

Finally, the tool also enables the XML exchange of metadata for the entire collection. The XML generated is an extension of the usual XML format generated by the *ImportExport* component that includes the specific metadata of the collection record, the specific metadata of each component, and the characterization of the aggregation relation used in this collection. Additionally, with this tool it is also possible to generate complete metadata descriptions of the components as if we had created each metadata record individually. The advantage of this approximation is the avoidance of metadata replication. Only a few metadata elements must be revised for each component and this is particularly relevant if the size of the collection is quite large (e.g. a collection composed of thousand of files).

5.3.4 Thesaurus Management

The *ThesaurusManagement* component is an enhanced thesaurus editor that has two main objectives: a basic management of thesauri according to the ISO norms for monolingual and multilingual thesauri (ISO 2788 (ISO, 1986) and ISO 5964 (ISO, 1985) respectively); and a second set of tools to enhance cross-discipline interoperability between different thesauri. These enhanced functionalities are based in the use of the WordNet lexical database and they mark the difference with respect to other thesaurus edition tools (a revision of thesaurus editors can be found at appendix C.2). Providing a WordNet interface, this tool implements the method for the disambiguation of thesauri presented in chapter 4. And thanks to this semantic disambiguation, the tool facilitates the automatic expansion of thesaurus terms with new terms from other thesauri having an equivalent meaning.

The tool has been developed in Java and it is deployed with two levels of operation: a simplified version which stores the thesaurus structure (only *BT*,*NT* relationships) on an Access 2000 database; and a complete version with full functionality that stores thesauri in an Oracle 9i database. The complete version takes advantage of the Oracle Intermedia Text package (CTX_THES) capabilities. This package implements the ISO 2788 norm for monolingual thesauri and also provides language translation relationships.

Figure 5.10 shows the sub-components (packages) of this tool. The functionality offered by this tool can be summarized as follows:

- *Basic Thesaurus Management.* This package facilitates the basic functionality for the edition and browsing of thesauri. Figure 5.11 displays the graphical user interface of this application. The visualization and browsing of thesaurus terms is possible with several graphical interface presentations and according to the language selected by the user (if exists translation).

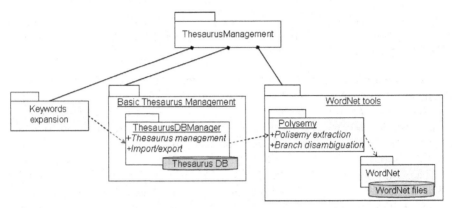

Fig. 5.10. Layered architecture of Thesaurus Tool

At present alphabetical and hierarchical presentations of thesauri are available. For instance, figure 5.11 shows how a user has displayed the UNESCO thesaurus (UNESCO, 1995) and the GEMET thesaurus (EEA, 2001), selecting different presentations (alphabetical presentation for UNESCO and hierarchical for GEMET) and different languages (English for UNESCO and German for GEMET). Additionally, to facilitate the discovery and visualization of terms, it is possible to perform "like" type searches. For example, the term *BIOCHEMISTRY* (or *BIOCHEMIE* in German) has been browsed in both thesauri, showing a different hierarchical path of terms in every case. Finally, the explicit relationships of a selected term are displayed on the right part of the *Thesaurus Viewer* window. There, users can edit the explicit relationships among concepts: synonym terms (SYN,USE), broader terms (BT), narrower terms (NT), related terms (RT), preferred terms (PT), scope notes (SN) and language translations (TR).

- *WordNet tools.* This package provides the interface to the WordNet lexical database. Additionally, subpackage *Polisemy* provides additional functionality to extract the senses of a term (or set of terms) and implements the disambiguation algorithm that was presented in chapter 4.
- *Keywords expansion.* This package facilitates the automatic detection of keywords that may be related to an initial set of keywords selected by a metadata creator.

The rest of this section will be devoted to the description of the enhanced capabilities of this thesaurus editor, i.e. the WordNet interface, the semantic disambiguation, and the keywords expansion functionality. Further details about the functionality of this tool can be found at (Nogueras-Iso et al., 2003a).

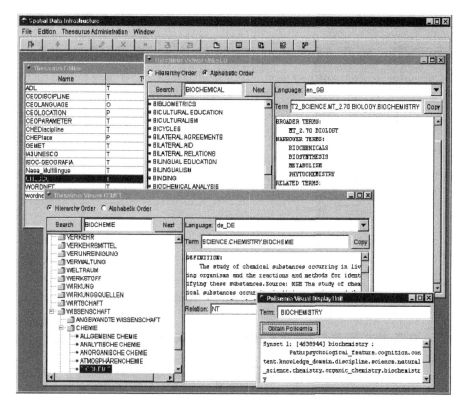

Fig. 5.11. Overview of the graphical user interface

WordNet interface

First of all, this tool allows the visualization of WordNet ontology, as if it were another thesaurus created by the tool (see figure 5.12). WordNet can be considered as an upper-level ontology which is structured in a hierarchy of synsets, where synsets are defined as set of synonyms representing a particular concept.

This functionality is provided by the Java *WordNet* package depicted in the architecture. This package facilitates the access to the libraries able to browse the lexical database. As the software of these libraries is implemented in C language and our application has been developed in Java, we had to implement the crosswalk that access to WordNet native libraries via JNI (Java Native Interface) and returns the information of the synsets in the same way as the information related with thesauri created by the tool.

Given that this tool provides access to WordNet, it also facilitates the possibility to find the senses of a term (single word or set of related words) in WordNet. This functionality is provided at low level by the package *Polisemy*, which was presented in the architecture. Given a term, this package looks up

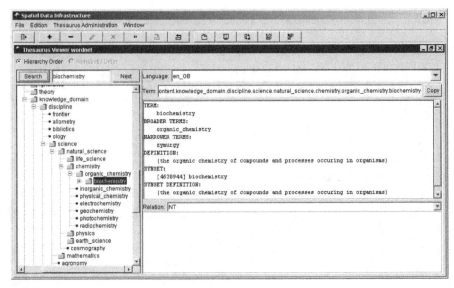

Fig. 5.12. Visualization of WordNet ontology

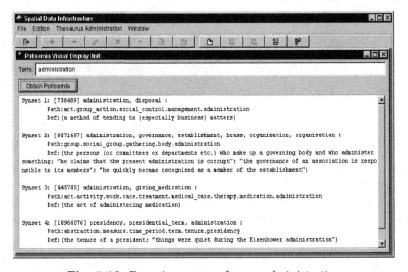

Fig. 5.13. Browsing senses of term *administration*

it in the WordNet database and extracts all the possible synsets. In case a
term is a compound term (more than one word) and is not directly included in
WordNet, the *Polisemy* component would extract all the synsets corresponding to each word in the compound term. Furthermore, this component uses
morphological techniques to reduce the number of not-found words and to
search the senses of adjectives which are associated with a noun. For instance,

given the adjective *administrative*, the component will look the synsets associated with *administration*. Figure 5.13 displays the window that facilitates the extraction of WordNet synsets given a term or phrase. In this case, figure 5.13 shows the senses of the polysemic term *administration*.

Semantic disambiguation of thesauri

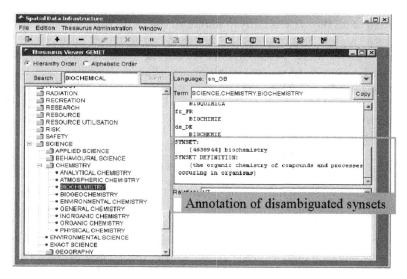

Fig. 5.14. Disambiguated senses are also shown by the thesaurus viewer

Whenever a user selects the option of importing a new thesaurus (from a text file in a pre-established format), the user is allowed to apply the semantic disambiguation of this thesaurus. The disambiguation algorithm (presented in chapter 4) associates each term in the thesaurus with its disambiguated sense (synset) in the WordNet upper-level ontology. The associated WordNet synsets are stored as a TR relationship (using $SYNSET$ as language). This TR relationship is used to indicate the translation of a term and in this case $SYNSET$ language it is interpreted as the disambiguation language. Once the disambiguation method is applied, the user is able to visualize the synset that was finally assigned (see fig. 5.14) to each term in the same way as other relationships. In fact, we could update manually the synset assigned to a thesaurus term.

Expansion of Keywords

Another application of thesauri disambiguation that has been incorporated within the tool is the expansion of keywords. Given a set of keywords belonging to an initial set of thesauri, this tool suggests a set of terms in a new thesaurus which share a similar sense with the ones selected initially. Metadata creators could profit from this tool to expand the keywords section of metadata records and enhance the description of resources. Additionally, this automatic expansion method could be also applied to enhance a user query.

The method to expand the keyword section is based on a basic routine which estimates the probability to expand an original set of keywords with a new term belonging to a new different thesaurus, not used in the original set. This basic routine is composed of two main steps. The first step is the collection of all the synsets corresponding to the terms, which were selected by metadata creator. As a result of this first step, we obtain an initial collection of synsets. And secondly, a comparison between the synsets of the new term and the initial collection of synsets is performed. This comparison consists of the computation of a reliability percentage for the new term, which is calculated as the number of synset coincidences divided by the number of synsets of the new term and multiplied by 99:

$$reliability = \frac{|synset\ matches\ of\ new\ term|}{|synsets\ of\ new\ term|} \times 99 \qquad (5.1)$$

The reason to use a final factor of 99 and not 100 in equation 5.1 is to obtain a maximum reliability percentage of 99 for automatically expanded terms, reserving uniquely a 100-reliability percentage for the terms which were originally selected by metadata creators. If this reliability percentage is greater than a *threshold* reliability percentage, which was defined previously by the user who performed the expansion, this new term is added.

Table 5.1. Manually introduced classifications

Thesaurus	Original term	Disambiguated synsets	Reliability
CEOPARAMETER	earth science → atmosphere	6270068	100
CEODISCIPLINE	weather & climate	7847974, 10413828	100

Synset id	Noun	Definition
6270068	atmosphere	(the mass of air surrounding the Earth; "there was great heat as the comet entered the atmosphere")
7847974	weather, weather condition, atmospheric condition	(the meteorological conditions: temperature and wind and clouds and precipitation; "they were hoping for good weather"; "every day we have weather conditions and yesterday was no exception")
10413828	climate,clime	(the weather in some location averaged over some long period of time)

As an example of this capability, the expansion of terms appearing in table 5.1 will be shown. These terms could correspond to the manual classification of a resource, which is included within the keywords section of a metadata record. The terms were selected from CEODISPLINE (a controlled list of 30 terms proposed to identify disciplines) and CEOPARAMETER (a controlled list of 1037 terms proposed to identify the types of features contained in a geospatial data resource) thesauri that are defined in (CEO, 1999).

Table 5.2. Terms automatically expanded with *thresshold=49*

Thesaurus	Expanded term	Synsets	Reliability
GEMET	atmosphere	**6270068**	$1/1 \times 99 = 99$
	climate	**10413828**	
	climate → weather	**7847974**	
	climate → weather → weather condition	**7847974**	
ADL-FTT	regions → climatic regions	**10413828**, 6359477	$1/2 \times 99 = 49.5$
NASA	atmospheric science	**6270068**, 4596663	$1/2 \times 99 = 49.5$
	atmospheric science → atmospheric tempera-ture	**6270068**, 3914851	

Table 5.2 shows the results of the expansion method for the input terms in table 5.1, all of them having a reliability value over 49. Summing up, 7 new terms were found, which belonged to three better structured thesauri: GEMET, the Alexandria Digital Library Feature Type Thesaurus (Hill, 2002) (ADL-FTT), and a list of thematic keywords proposed by the NASA Global Change Master Directory (GCMD) project[2].

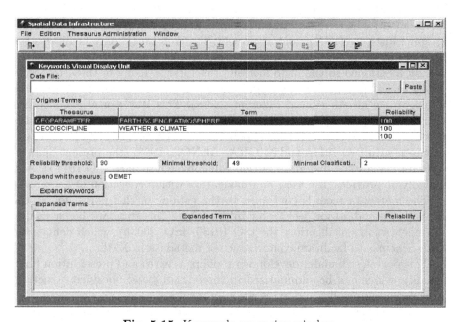

Fig. 5.15. Keywords expansion window

Figure 5.15 displays the window that facilitates the expansion of keywords. This window assists the work of metadata creators by suggesting similar terms

[2] Draft geospatial thematic keywords from the NASA GCMD in short and long format for CSDGM of FGDC. Available at http://www.fgdc.gov/clearinghouse/reference/refmat.html.

from other thesauri. Otherwise the user should filter the terms of new thesauri on his own, a time-consuming task when thesauri contain thousand of terms.

5.4 The Web Portal of a Spatial Data Infrastructure

This section illustrates an example of a customized search interface that makes profit of the enhanced functionalities of the catalog server component for collection management and metadata interoperability.

As a use case, this section presents the portal that has been developed at the University of Zaragoza for the *Spanish Spatial Data Infrastructure* (Infraestructura de Datos Espaciales Española, IDEE). Although several projects and studies (Bañares et al., 2001; Bernabé et al., 2001, 2002; Gould et al., 2002; Nogueras et al., 2001a) had motivated the necessity for the construction of a Spanish SDI, it was not until the end of 2003 that it appeared the first official initiative for the construction of the infrastructure (Béjar et al., 2004). Therefore, this infrastructure is still (April 2004) in its initial phase of development. The infrastructure is coordinated by the Spanish National Geographic Institute (IGN)[3] and at present, the Web Portal only facilitates the discovery of the geographic resources produced or owned by this institute. As already mentioned in the introduction of chapter 2, many of these resources are organized in collections (cartographic series arisen from the fragmentation of geographic information at different scales). This Web portal is concerned with the concepts presented in this book in two main aspects. Firstly, it takes into account the problem of interoperability as it presents metadata in conformance with ISO 19115 and MIGRA (AENOR, 1998) (the present Spanish norm for geographic metadata, which will converge soon to ISO 19115). And secondly, it provides discovery and navigation within collection of resources.

The metadata records contained in the catalog of this portal follow the ISO 19115 standard for geographic information. This standard has a related implementation specification, the ISO 19139 (ISO, 2003b), which defines the XML-Schemas to facilitate conformance of metadata in XML.

This portal, still under development, offers a search and presentation functionality, which can be summarized in three main steps: query construction, presentation of results, and exploration of results.

Firstly, the client must specify a query restriction. For instance, in figure 5.16 (left side) the user has specified a restriction to retrieve the datasets covering the northeastern part of Spain. Additionally, it can be observed that this search interface does not overwhelm the user with an infinite list of fields to specify restrictions on all the possible metadata elements. It is believed that search interfaces uniquely based on the direct association of metadata elements and values result too complex for a not experienced user. Thus, this search interface is based on an increase in the level of abstraction as

---

[3] http://www.ign.es/

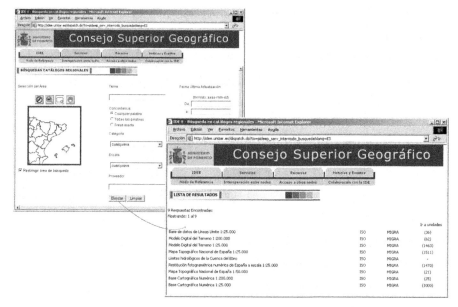

Fig. 5.16. Specifying a restriction(left) and browsing the results(right)

explained in (Cantán et al., 2003). Apart from the spatial extent restriction, it only offers other four search fields (theme, topic category, provider and time extent) for specifying restrictions, which are internally extended to all the metadata elements related to these abstract concepts.

In a second step, a list of aggregated results is presented to the user (see right side of figure 5.16). These results are categorized in three types: datasets without further decompositions that verify the restriction; collections whose metadata verify the user restrictions; and collections that are returned because more than two of their components verify individually the restriction specified by the user. Thanks to this aggregated presentation of results, the user is not overwhelmed with thousand of components of the same collection.

And thirdly, the user may explore the results in more detail. Here, the user has three main options.

- The user may refine the query adding more restrictions.
- The user may browse the complete metadata descriptions of the records returned in conformance with ISO 19115 or MIGRA format. Figure 5.17 (left side) shows an example of these metadata descriptions.
- And alternatively, in the case of having results that are collections, it is possible to click on the number of components (it appears between parentheses on the right side of the title) that verify the restriction, and browse the list of these components. Figure 5.17 (right side) shows an example of this detailed refinement through the components of the collection. And as

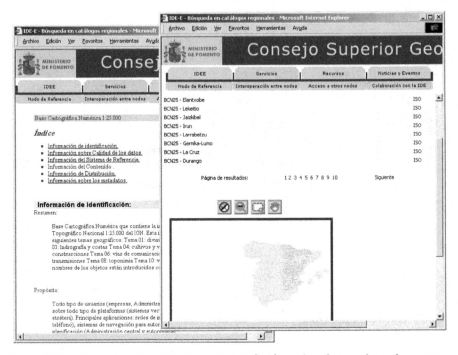

Fig. 5.17. Browsing ISO 19115 metadata(left) and refining through components(right)

it can be observed, apart from the typical list of component titles, the user is also provided with a map view displaying the area covered by the components that verify the restriction in the collection. This map is obtained through a remote Web Map Server (Beaujardière, 2002) which overlays two layers: one layer corresponding to the spatial distribution pattern of the collection, and a second layer which correspond to the tiles of the components verifying the restriction specified by the user.

5.5 Conclusions and future work

This chapter has demonstrated that the concepts presented in previous chapters have contributed to the development of the components of a spatial data infrastructure with enhanced capabilities.

Firstly, the combination of the proposed solution for collection management, the crosswalks between metadata standards and the use of information retrieval strategies have enabled the development of a flexible catalog services component. Apart from the use of the Metadata Knowledge Base presented in chapter 2, the crosswalks created through the process explained in chapter 3 have allowed the construction of a catalog that can be considered as metadata-interoperable. It admits restrictions over elements belonging to different standards and returns results formatted to the standard required by the user. Additionally, the concept-based retrieval strategy 4 has been integrated within the catalog services to provide an additional factor to compute the overall relevance of metadata records with respect to the user query.

Secondly, this chapter has presented a metadata edition tool, called *CatMDEdit*, that provides additional functionalities with respect to the existent metadata editors. Apart from the edition and exchange of metadata according to different standards, it enables a unified edition of collections of related resources and includes an enhanced tool for the selection of thesaurus terms. Furthermore, it must be remarked that this tool has been deployed for several projects in Spain. It is used at the Environment Department of the government of Galicia region in the construction of its spatial data infrastructure (Béjar et al., 2003a). The Ebro River Basin Authority (CHE) also uses this tool for cataloguing their geographic information (Arqued-Esquía et al., 2001). This Spanish public institution is in charge of the physical and administrative management of the hydrographical basin of the Ebro River. And last, it is freely distributed through the Web Portal of the incipient Spanish Spatial Data Infrastructure, which is coordinated by the Spanish National Geographic Institute (IGN).

Thirdly, this chapter has shown how the Web Portals of a spatial data infrastructure make use of catalog services to offer a customized search interface of their resources. In particular, this chapter has presented the functionality offered by the Web Portal of the Spanish Spatial Data Infrastructure. The main attraction of this customized search interface is the possibility of presenting collection aggregated results instead of overwhelming the user with thousand of records describing the components of the same collection.

Finally, the future lines of the applications presented in this chapter will be directed to the stabilization and improvement of the additional capabilities that have been described. In the case of the metadata edition tool, the work will continue on facilitating the support of other types of collections (e.g., temporal, spatio-temporal, thematic, etc.) and the management of recursive levels of aggregations, i.e. collections whose components also aggregate a set of datasets. The catalog services already give support for these types of collections but an appropriate interface must be given to metadata creators in order to facilitate their work. In the case of customized Web search interfaces, further possibilities must be studied to facilitate the discovery of resources. For instance, the user could be provided with enhanced capabilities to: save query results as new collections; link to related collections that also include the initial resource that the user found; or, in the case of spatial collections, find components in different collections using the same spatial distribution pattern that may seamlessly overlayed. Finally, with respect to the thesaurus management software, the creation of a Thesaurus Web Service has been planned. This service would provide similar functionality to the one described within the metadata edition tool (e.g., on-line thesaurus browsing, WordNet polysemy extraction or keywords expansion), but this time as a member of the on-line services offered by the SDI. The advantage of having this service on-line would be that Web search interfaces could incorporate more sophisticated topic-based searches and recommend the user a more appropriate vocabulary for their queries.

6

Conclusions and future work

Spatial data infrastructures provide the framework for the optimization of the creation, maintenance and distribution of geographic information at different organization levels (e.g., regional, national, or global level) and involving both public and private institutions. From a technical point of view, the development of such an infrastructure requires the combination of technologies coming from a background in multiple disciplines. Overall, among these disciplines the experience of geographic information systems and digital libraries is particularly relevant. But perhaps the most distinctive features of these infrastructures are the use of catalogs and metadata, which are used to interconnect the data and services offered by a spatial data infrastructure. In fact, descriptions of data and services are closely related. The metadata describing the geospatial data holdings can be used to derive the metadata describing the capabilities of the services (Web Map Server, Web Feature Server, etc.) that provide access to these holdings. And similarly, metadata describing services are used as entries to the services catalogs (registries) that publicize the range of services offered by the spatial data infrastructure.

The work presented in this book has been focused on the technologies and methodologies that can encourage the development of spatial data infrastructures by means of a better utilization of metadata. As starting point for the work presented in this work, three main problems were identified as hindering the correct use of metadata. The first problem was concerned with the large number and high volume of geographic resources, which increases the complexity of cataloguing them correctly. However, it is common to find that, at least, it is possible to identify groups of related resources which could be catalogued together. These groups of related resources are commonly called collections and they usually arise as a result of the fragmentation of geographic resources into datasets of manageable size and similar scale. With respect to the cataloguing purposes, the most significant of feature of such collections is that their components share a high percentage of metadata and that the organization of data holdings according to the hierarchical structure of collections would facilitate the discovery and access services of a spatial data infrastructure. The other two problems making the use of metadata more difficult were related to the heterogeneity of metadata. As mentioned in (Méndez-Rodriguez, 2002) (pp. 216), metadata systems differ in two main aspects: structure and content. The first aspect is concerned with the diversity of metadata standards and schemas. Along the last decade many organizations (standardization bodies, software vendors, ...) started different initiatives for the definition of metadata standards with the goal of describing the features of different types of media objects and promoting the common understanding within a community of users. However, despite the initial intention of common understanding, the diversity of initiatives originated an undesired effect of heterogeneity. The second aspect of this metadata heterogeneity problem is the content heterogeneity. By content heterogeneity it is meant the problem of identifying that the values of two metadata elements are meaning the same concept despite using different terms. When the values of metadata elements are constrained to a predefined list of values, there is no doubt. But, if the domain of a metadata element is free-text data, possible misunderstandings may appear. This situation is usually minimized by the use of a normalized vocabulary (e.g., thesauri) but, despite this, the catalog discovery services should not be uniquely implemented as a simple word matching between the user queries and the metadata records stored in the catalog. On the opposite, catalog services should consider the use of information retrieval strategies, which are concerned more with retrieving information about a subject than retrieving data which satisfies exactly a given query.

Therefore, the goal of the work presented in this book has been to increment the capacities of metadata catalogs in three main research lines: the support for the management of nested collections, the interoperability among different metadata standards, and the incorporation of information retrieval techniques.

As regards the management of nested collections, this work has proposed the design of a catalog system that is based on the use of a Metadata Knowledge Base component. The main features of this knowledge base component are the use of XML technologies and the improvement in the expressive power of the aggregation relations that define the components of a collection. This knowledge representation approach, partially presented in (Nogueras-Iso et al., 2004f), can provide great benefits for the construction of metadata cataloguing systems, either applied within the context of a spatial data infrastructure or within the more generic context of digital libraries. And the main conclusions that can be obtained from this approach are:

- The avoidance of the redundancy in the metadata creation process. Thanks to this approach, metadata is only maintained in one place and inferred whenever it is needed. The expressive characterization of aggregation relations facilitates the automatic inference of meta-information for both components and collections metadata records. A general characteristic of the components of a collection is that they share a high percentage of meta-information and thanks to this metadata inference; it is possible to segregate the meta-information at the appropriate level of commonality or specificity, thus avoiding redundancy of information.

- Secondly, this approach facilitates the supervision of metadata creation process. This knowledge representation enables the specification of patterns that the components of the collection should follow and thus, the status of cataloguing will be supervised by comparison with the patterns. For instance, in the case of a spatial collection, it is possible to overlap the spatial pattern grid (the division of tiles for a specific scale) and the layer formed by the bounding boxes of the components already catalogued. Additionally, the metadata for the components of a spatial component could be graphically edited and facilitated by this spatial pattern (in the form of a coverage).

- The use of this enhanced catalog enables the discovery and presentation of metadata records at an aggregated or disaggregated level on user demand. The knowledge base can deduce whether an initial set of metadata results are describing components of the same collection, i.e. the knowledge base could find the metadata record that subsumes the initial results in the ascending whole-part hierarchy. Thanks to this, the system can present only an aggregated view of query results to the user in a first step, and a detailed view of the components metadata in a second step. Furthermore, for this second filtering the user can make profit of the collection pattern that defines the distribution of components.

- And last, the unified description of collections and components may also help to generalize software for access and visualization of aggregated resources. For instance, an enhanced implementation of Web Map Servers could make profit of the modeled aggregation relations to display automatically aggregations of datasets that form part of the same collection.

Concerning the interoperability between different metadata standards, this book has presented a methodology to carry out the construction of a series of crosswalks that enable the conversion between different metadata schemas. Two main reasons motivate the creation of crosswalks: the convergence towards international standards and the reusability of resources across different domains. On one hand, the standardization initiatives within each application domain have usually converged to an international standard but the legacy metadata (the work done in the past) cannot be directly thrown away. And on the other hand, the search of resources across different domains is still needed. Although digital libraries may be specialized on particular types of resources and use specific metadata for such resources, they are also asked to provide general descriptions of their resources for the sake of interoperability. The crosswalk creation process presented in this book consists of a series of steps that gradually incorporate fine grained details about the source-to-target mapping until the full crosswalk is finished. And from this crosswalk creation process, which has been partially presented in several works (Anaya et al.,

2002; Lacasta et al., 2003; Nogueras-Iso et al., 2004d), it can be concluded that:

- Thanks to this process, it is possible to establish a semi-formalized method that implies a rigorous specification of standards and transformations, which minimizes the possible loss of information. Furthermore, this work has also proposed (within the implementation phase of this process) the design of a semi-automatic tool for the implementation of crosswalks, which alleviates the hard and error-prone task of coding XSLT instructions, the technology selected for the translation of metadata records encoded in XML (de facto standard for exchange format).
- Additionally, it must be remarked that this process is context free and thus, it can be applied to transform metadata in any domain context or even to transform source and metadata schemas from different domains.

And with respect to the incorporation of information retrieval strategies, this book has presented an information retrieval strategy that facilitates the metadata content interoperability between metadata repositories using heterogeneous vocabularies. The participation of several organizations, probably ranging from different application domains, may be the cause of this content heterogeneity. However, this heterogeneity may be simply motivated by a different point of view of several metadata creators too. To overcome this heterogeneity, this work has presented an unsupervised technique for the disambiguation of thesaurus terms (our selected vocabulary) in light of their surrounding terms and with reference to an external upper level ontology, WordNet in this particular case. This method takes advantage of the thesaurus hierarchical structure (broader and narrower terms), which is used as the word context for a voting algorithm to find the closest sense. And thanks to this disambiguation, this book has proposed an information retrieval strategy adapting a classic information retrieval model to the context of a digital library, understood as a catalog holding metadata records. The information retrieval model has made use of the homogeneous indexing provided by the WordNet synsets, which are the disambiguated senses of each thesaurus term included within the metadata records. Apart from being presented in this book, the adaptation of this retrieval model to the context of metadata catalogs has been also introduced in (Nogueras-Iso et al., 2003b, 2004a) [1]. And the main conclusions that have been obtained could be summarized as follows:

- As concerns the disambiguation method, the main disadvantage is that it may not be adequate for the semantic disambiguation of very specific

---

[1] (Nogueras-Iso et al., 2004a) is an extended version of (Nogueras-Iso et al., 2003b) that was accepted for a special volume of the LNAI series. The extended version incorporates experiments for the evaluation of the information retrieval efficiency of our approach (including the comparison with similar approaches), thus allowing the verification of the initial proposals.

domain ontologies because WordNet lacks for domain-specific terminology. Nevertheless, the intention of this work is to approximate as much as possible the terms used in metadata records and the concepts extracted from "general-purpose" queries. Furthermore, this disambiguation algorithm provides the basis for a wide range of interesting applications.

- And regarding the information retrieval model, the applicability of the vector-space model has been explored. In more heterogeneous contexts, other retrieval models, such as probabilistic or neural-net based models, would work probably better. However, in this context of metadata catalogs, the own metadata records are the summary of the desired resource and a simple model may provide satisfactory results. Finally, as far as the indexing of metadata records is concerned, it is worth mentioning that apart from collecting the synsets associated with a thesaurus term, the indexing method also tested the possibility of including other synsets associated with other related terms in the thesaurus hierarchy. In particular, the inclusion of synsets associated with the broader terms in the thesaurus hierarchy was proposed. This modification was based on the assumption that metadata records represented by these synsets (from broader terms) are still semantically close to queries including the broader concept. This expansion could have been also continued with the synsets associated with other related terms. However, it is believed that resources containing concepts at distance two or more from the initial concept expressed in the user query are not relevant to the user information needs (Clark et al., 2000).

The viability of the aforementioned proposals has been always tested within the context of spatial data infrastructures. Firstly, the collection management support has been illustrated with examples of geographic collections (temporal series, mosaics of images, etc.), which are very frequent in this context. Secondly, the process for the construction of crosswalks was used to obtain a series of crosswalks between the most important geographic metadata standards, and also providing interoperation with the general purpose Dublin Core standard. Thirdly, the information retrieval strategy was tested with a collection of metadata records describing geographic resources and this strategy was compared with a typical word-based retrieval system. The first experiments showed that both strategies are comparable in terms of precision and that our proposed strategy improves the recall measures, i.e. it discovers resources related with queries despite not using the same words. And finally, all these proposals have contributed to the development of enhanced components of spatial data infrastructures, mainly those concerned with the development of catalogs and their client applications. The addition of crosswalks and the concept-based retrieval strategy together with the use of a metadata knowledge base has enabled the development of a robust catalog services component which can be considered as metadata-interoperable, supports the management of nested collections and takes into account possible concept matches. And around this catalog services component, a set of client applications such as metadata editors or the Web portals of a spatial data infrastructure have exploited its enhanced capabilities. In the latter case, the search interfaces are particularly benefited with the possibility of presenting collection aggregated results instead of overwhelming the user with thousand of records describing the components of the same collection.

However, all the concepts and ideas that have been presented in the work presented in this book are perfectly applicable to more generic contexts. On one hand, all the proposals for the improvement in the utilization of metadata and its related components can be extended to the more general context of digital libraries. And on the other hand, each individual proposal is also applicable in other fields. Firstly, the interoperability between metadata standards can be generalized to the problem of heterogeneity between semi-structured data sources, e.g. those sources represented in XML, or even to the problem of heterogeneous databases in case of abstracting us from the use of XML. And in addition to this, the information retrieval strategies could be also extensible to the information retrieval of any type of document containing an identifiable set of keywords, e.g. keywords of papers in journals, or the table of topics that appears sometimes in the back of a scientific book.

Finally, along the elaboration of the work presented in this book, it has been observed that several issues could constitute the research lines representing the continuation of this work. They are the following:

• The support for new types of relations in the Metadata Knowledge Base that provides the base for the catalog services. Apart from modeling the aggregation relation, other types of relations could be also benefited from the advantages that the metadata knowledge base approach provides: automatic metadata inference mechanisms, generation of statistics, navigation through relationships, and so on. In this sense, we have already detected several relation types in the context of Geographic Information. One of these relations could be identified as a *version* relation. This relation reflects the association established between a set of source datasets and a dataset that has been derived from these source datasets. This relation could be even specialized depending on the type of transformation that is performed over the source datasets. Some examples of these specializations are the following: coordinate projection transformations, spatial representation transformation, operations on themes. Another important type of relations could be entitled as a *revision* relation. This relation reflects the association between a source dataset and the datasets derived from the previous one by correcting or revising some attributes values. The metadata record describing the new resource is practically identical to the original one except for the addition of details in the *data quality* section or the modification of dates (e.g., *temporal extent, publication date*, etc.). Another type of relation, which must not confused with a *revision* relation, is the *format* relation. This relation reflects the association between a source dataset and the datasets derived from the previous one by delivering the same contents but in a different format. Once again, the metadata record describing the new resource is practically identical to the original one except for some additional details in the data quality section and the details of this new format in the *distribution information* section. And last, we have also identified a special type of relations identified as *high-level aggregation* relations. Apart from giving support for collections where all the metadata records describing the components reside in a local catalog, we may encounter that geographic resources and their metadata are distributed at different nodes of a spatial data infrastructure (e.g., risk management scenarios in cross-border areas). Furthermore, the metadata records describing each individual resource may be distributed across the different geographic data catalogs. In this case, it would be interesting to model a high-level aggregation relation pointing at the metadata records in each remote catalog.

- A more automated assistance in the process of crosswalk creation. Future lines of the process of crosswalk construction should be oriented to the design of a CASE tool assisting this process. Despite having described a semi-formal method, this process is still error-prone if not done with enough thoroughness. Thus, perhaps the main challenges of this CASE tool will be: provide help in the harmonized description of standards; facilitate the semantic mapping between the source and target standards; and the further automation in the creation of stylesheets. Firstly, XML-Schemas and DTDs could be used to generate as much as possible harmonized descriptions of the source and target standards. Secondly, an initial semantic mapping could be automatically proposed by means of the linguistic analysis of element terms using dictionaries and lexical ontologies. That is to say, the element terms from both standards would be disambiguated against an upper-level ontology in order to recognize possible links. Obviously this linguistic mapping will not be exact but it will probably detect some obvious mappings that can save time of the user. And finally, the automatic creation of XSL documents could be improved. This work has presented a solution to generate automatically the initial versions of stylesheets. However, most of the additional transformation rules must be still hand-coded. Most of these problems are highly context-specific and it is difficult to find general patterns applicable in different crosswalks. Thus, further research must be done in the categorization and specification of these rules to facilitate their automatic translation into a series of XSLT instructions.
- Extensions of the semantic disambiguation method. The method for the disambiguation of thesauri presented in this book has only addressed the interconnection of English thesauri. One of the remaining tasks of this disambiguation method is to enable the interconnection of thesauri in other languages. In this sense, the use of EuroWordNet, which connects WordNets in different languages, is a promising approach to allow cross-language disambiguation. This multilingual disambiguation of thesauri will

contribute to the development of multilingual catalog search services (see (Nogueras-Iso et al., 2004e) as a first approximation). Additionally, apart from using the disambiguation algorithm to disambiguate thesaurus terms, this algorithm could be also applied to other types of resources having a hierarchical structure. The disambiguation method uses the hierarchical structure of a thesaurus (hierarchy of broader and narrower terms) as the context for the disambiguation. And using this philosophy, the disambiguation method could be used, for instance, to disambiguate the content of XML documents making profit of the hierarchical structure of XML elements.

- Further improvements in the information retrieval model. An improvement in the computation of the weight of each index term would be to consider the importance of the thesaurus (i.e., a measure concerned with the specificity, size or maturity), to which the terms in the keyword section belong. Another improvement in the method could be a better representation of the user query. Apart from the synsets related with words contained in the query phrase, the ancestors of these synsets in WordNet hierarchy could be also considered. In this way, the information retrieval system could return, at least, metadata records referencing synsets in the ancestor hierarchy of the query synsets. Furthermore, the words contained in the definition of the WordNet synsets could be used to expand the query formulated by the user. Additionally, it must be mentioned that this retrieval method could be extended by indexing other disambiguated metadata elements such as the *title* or *abstract*.

- The continuation in the development of enhanced components to be integrated within a spatial data infrastructure. From a more technical point of view, the development of several components is still open:

  - In the case of the metadata edition tool, the work will continue on facilitating the support of other types of collections (e.g., temporal, spatio-temporal, thematic, etc.) and the management of recursive levels of aggregations, i.e. collections whose component also aggregate a set of datasets. The catalog services already give support for these types of collections but an appropriate interface must be given to metadata creators in order to facilitate their work.

  - In the case of customized Web search interfaces, further possibilities must be studied to facilitate the discovery of resources. For instance, the user could be provided with enhanced capabilities to: save query results as new collections; link to related collections that also include the initial resource that the user found; or, in the case of spatial collections, find components in different collections using the same spatial distribution pattern that may seamlessly overlayed.

  - And with respect to the thesaurus management software, the creation of a Thesaurus Web Service has been planned. This service would provide similar functionality to the one described within the metadata edition tool (e.g., on-line thesaurus browsing, WordNet polysemy extraction or keywords expansion), but this time as a member of the on-line services offered by the SDI. The advantage of having this service on-line would be that Web search interfaces could incorporate more sophisticated topic-based searches and recommend the user a more appropriate vocabulary for their queries.

A

Collections

A.1 Consistency of the metadata model

Before designing the Knowledge Base presented in section 2.4 to tackle the problem of cataloguing collections, it was studied how metadata records could be synthesized into a minimized model. That is to say, having an initial collection scenario where a metadata record had been created to describe individually (and with consequent redundancies) each component of the collection and the entire collection, we wanted to find a way to transform these initial records into a set of minimized records. Furthermore, our intention was to demonstrate that this transformation function was biyective, i.e. there was a mapping 1:1 between the original and the minimized model. Thus, if a system contained this minimized model, it would be possible to restore the original model when needed.

In order to find out this possible transformation function assuring no loss of information and the possibility of reversibility, we opted for considering metadata records as Abstract Data Types (ADT). There are numerous works in the literature that use algebraic specifications of ADTs, and in general formal specifications, as design tools (Guttag and Horning, 1978, 1980; Horebeek and Lewi, 1989). On one hand, the operations defined for this ADT could facilitate the work of finding this transformation function. And on the other hand, the algebraic specification of this operation could demonstrate the reversibility of this transformation/minimization of the original metadata model.

For the definition of the metadata record ADT we have considered that a metadata record could be defined as a flattened list of ordered elements. This does not constrain the generality of the definition because each metadata element could be a complex structure if necessary. Figure A.1 shows the specification of a metadata record using the algebraic data type language ACT ONE (part of LOTOS (Turner, 1993)).

In this specification, the *sorts* and *opns* parts declare the sorts and specify the operators (constants and nullary operators) along with the signature for the type. The main operations defined for this ADT are *generalization*,

1: **type** $metadataRecord$ **is**

2: **formalsorts** $metadataElement, natural$

3: **sorts** $metadaRecord$

4: **opns**

5: $emptyRecord :\rightarrow metadataRecord$ (* empty record *)

6: $setEl : metadataRecord, natural, metadataElement \rightarrow metadataRecord$ (* obtains the value of an element *)

7: $getEl : metadataRecord, natural \rightarrow metadataElement$ (* sets the value of an element *)

8: $\_ \bigtriangledown \_ : metadataRecord, metadataRecord \rightarrow metadataRecord$ (* extension *)

9: $\_ \triangle \_ : metadataRecord, metadataRecord \rightarrow metadataElement$ (* generalization *)

10: $\_ - \_ : metadataRecord, metadataRecord \rightarrow metadataRecord$ (* subtraction *)

11: **eqns**

12: **forall** $a, b, c : metadataRecord$

13: **ofsort** $metadataRecord$

14: $(a \triangle b) \triangle c = a \triangle (b \triangle c)$; (* generalization associative property *)

15: $a \triangle b = b \triangle a$; (* generalization commutative property *)

16: $a = (a - b) \bigtriangledown (a \triangle b)$; (* equivalence 1 *)

17: $a - (a \triangle b) = a - b$; (* equivalence 2 *)

18: (* .. Omitted .. *)

19: **endType**

Fig. A.1. The $metadataRecord$ ADT

subtraction and *extension*, which are represented by the symbols '$\triangle$', '-' and '$\bigtriangledown$' respectively. The axioms in the *eqns* section clarify the semantic of these operators. Informally, the intention of these operations is the following:

- The *generalization* operation should be used to obtain a new metadata record that contains the common metadata information of a set of metadata records.
- The *subtraction* operation between two records a and b should be used to obtain a new metadata record that discards from a the common metainformation that shares with b.
- And the *extension* operation between two records a and b should be used to obtain a new record that is the extension (union) of records a and b.

Figure A.2 shows the transformation between an original metadata scenario and a minimized scenario by means of *subtraction, generalization* and *extension* operations. This figure shows the same notation that was already used for figure 2.8 in section 2.3: MD_is and $MD\_Collection$ are the original metadata records describing the components and the collection; and $MDS\_collection$ and MDS_is are the records of the minimized model stored for the collection and each component. This transformation has the following features: the generalization of MD_is and $MD\_Collection$ generates a *common metadata* record (see later the comment about the subdivision of this *common metadata*); the subtraction of MD_i and *common metadata* generates the MDS_is; and the subtraction of $MD\_Collection$ and *common metadata* generates the *collection-specific metadata* compartment.

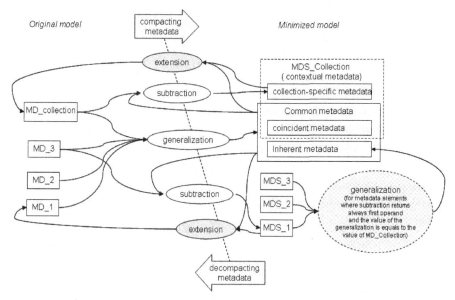

Fig. A.2. Transformation between original and minimized scenario

Theorem: $e_i = s_i \bigtriangledown g$ where

e_i is a metadata record

g is the generalization of a set of records including e_i: $g = (e_1 \bigtriangleup e_2 \bigtriangleup e_3... \bigtriangleup e_n)$

and s_i is the not common part of e_i: $s_i = e_i - g$

Proof:

$s_i \bigtriangledown g = (e_i - (e_1 \bigtriangleup e_2 \bigtriangleup e_3... \bigtriangleup e_n)) \bigtriangledown (e1 \bigtriangleup e2 \bigtriangleup e3... \bigtriangleup en) =$

(* aplying axioms in lines 14 and 15 of specification *)

$(e_i - (e_i \bigtriangleup (e_1 \bigtriangleup e_2 \bigtriangleup .. \bigtriangleup e_{i-1} \bigtriangleup e_{i+1} \bigtriangleup .. \bigtriangleup e_n))) \bigtriangledown (e_i \bigtriangleup (e_1 \bigtriangleup e_2 \bigtriangleup .. \bigtriangleup e_{i-1} \bigtriangleup e_{i+1} \bigtriangleup .. \bigtriangleup e_n)) =$

(* applying axiom in line 17 *)

$(e_i - (e_1 \bigtriangleup e_2 \bigtriangleup .. \bigtriangleup e_{i-1} \bigtriangleup e_{i+1} \bigtriangleup .. \bigtriangleup e_n)) \bigtriangledown (e_i \bigtriangleup (e_1 \bigtriangleup e_2 \bigtriangleup .. \bigtriangleup e_{i-1} \bigtriangleup e_{i+1} \bigtriangleup .. \bigtriangleup e_n)) =$

(* applying axiom in line 16 *)

e_i

Fig. A.3. Derived equation (theorem) from metadata record ADT

The question now is to check whether this transformation is reversible. That is to say, we should prove that the extension between *collection-specific metadata* and *common metadata* obtains again *MD_Collection*; and that the extension between *MDS_i* and *common metadata* obtains again *MD_i*. The answer is that given an implementation of the previous metadata record ADT, we could affirm that the previous hypothesis of the reversible transformation is possible. This is proven by a theorem (property) derived from the axioms in the *eqns* section of the ADT. This theorem could be stated as follows: "having the common meta-information of a set of metadata records (i.e., the generalization of this set of metadata records) and the non-common of each record (subtraction of each initial record and the common metainformation), it is possible to reconstruct every original metadata record". Figure A.3 defines more formally this theorem and its validity through the successive application of axioms.

1: **type** *metadataElement* **is**
2: **formalsorts** *dataType*
3: **sorts** *metadataElement*
4: **opns**
5: *emptyElement* :→ *metadataElement* (* empty element *)
6: *setValue* : *dataType* → *metadataElement*
7: *getValue* : *metadataElement* → *dataType*
8: _ ▽ _ : *metadataElement, metadataElement* → *metadataElement* (* extension *)
9: _ △ _ : *metadataElement, metadataElement* → *metadataElement* (* generalization *)
10: _ − _ : *metadataElement, metadataElement* → *metadataElement* (* subtraction *)
11: **eqns**
12: **forall** a, b, c : *metadataElement*
13: **ofsort** *metadataElement*
14: $(a \triangle b) \triangle c = a \triangle (b \triangle c)$; (* generalization associative property *)
15: $a \triangle b = b \triangle a$; (* generalization commutative property *)
16: $a = (a - b) \triangledown (a \triangle b)$; (* equivalence 1 *)
17: $a - (a \triangle b) = a - b$; (* equivalence 2 *)
18: (* .. Omitted .. *)
19: **endType**

Fig. A.4. The *metadataElement* ADT

An implementation of this metadata record ADT should be based on the existence of a metadata element ADT with similar operations and semantics. Figure A.4 shows the algebraic specification of the metadata element ADT. And based on this ADT, table A.1 presents the implementation of a metadata record.

Therefore, in order to enable this reversible transformation we should identify possible implementations of the metadata element ADT. These implementations usually depend on the data type of the element value. Some of the most typical implementations that we have identified are:

- The 'default' type implementation (it is shown in table A.2). This implementation is very general a does not bother very much about the *dataType* of the element value. The only restrictions on the *dataType* are the following: there is an equals (=) operator, and the elements can be assigned a *null* value. With respect to minimization in figure A.2, the idea behind this implementation is to detect common values of elements in *MD_is* and *MD_Collection* records. If all these values are equal, a unique value will be stored in the *common metadata* record. Otherwise, the original values are stored in *MDS_is* and *collection-specific metadata*.

Table A.1. The implementation of the *metadataRecord* ADT

Implementation of operations

Operation	Implementation
emptyRecord	A metadata record that consist of a list of empty elements.
getEl(a, n)	Typical operation in lists
setEl(a, n, value)	Typical operation in lists
$a \bigtriangledown b$	$\{getEl(a,1) \ \bigtriangledown \ getEl(b,1), getEl(a,2) \ \bigtriangledown \ getEl(b,2), \dots, getEl(a,n) \ \bigtriangledown \ getEl(b,n)\}$
$a \bigtriangleup b$	$\{getEl(a,1) \ \bigtriangleup \ getEl(b,1), getEl(a,2) \ \bigtriangleup \ getEl(b,2), \dots, getEl(a,n) \ \bigtriangleup \ getEl(b,n)\}$
$a - b$	$\{getEl(a,1) \ - \ getEl(b,1), getEl(a,2) \ - \ getEl(b,2), \dots, getEl(a,n) \ - \ getEl(b,n)\}$

Proof of the equations for this implementation of operations

Equation	Proof
$a \bigtriangleup b = b \bigtriangleup a$	This is derived from the 'generalization commutative property' of elements. $a \bigtriangleup b =$ $\{getEl(a,1) \bigtriangleup getEl(b,1), getEl(a,2) \bigtriangleup getEl(b,2), \dots, getEl(a,n) \bigtriangleup getEl(b,n)\} =$ $\{getEl(b,1) \bigtriangleup getEl(a,1), getEl(b,2) \bigtriangleup getEl(a,2), \dots, getEl(b,n) \bigtriangleup getEl(a,n)\} =$ $b \bigtriangleup a$
$(a \bigtriangleup b) \bigtriangleup c = a \bigtriangleup (b \bigtriangleup c)$	This is derived from the 'generalization associative property' of elements.
$a = (a - b) \bigtriangledown (a \bigtriangleup b)$	This is derived from the 'equivalence 1' axiom of elements.
$a - (a \bigtriangleup b) = a - b$	This is derived from the 'equivalence 2' axiom of elements.

Table A.2. The 'default' type implementation of the *metadataElement* ADT

Implementation of operations

Operation	Implementation
emptyElement	$setValue(null)$
getValue(a)	Obtain the value.
setValue(a, value)	Set the value.
$a \bigtriangledown b$	**if** $getValue(a) = null$ **then return** $setValue(getValue(b))$ **else return** $setValue(getValue(a))$
$a \bigtriangleup b$	**if** $getValue(a) = getValue(b)$ **then return** $setValue(getValue(a))$ **else return** $setValue(null)$
$a - b$	**if** $getValue(a) = getValue(b)$ **then return** $setValue(null)$ **else return** $setValue(getValue(a))$

Proof of the equations for this implementation of operations

Equation	Proof
$a \bigtriangleup b = b \bigtriangleup a$	Derived from implementation. The order of operands does not have any effect in the result.
$(a \bigtriangleup b) \bigtriangleup c = a \bigtriangleup (b \bigtriangleup c)$	If the values of a, b and c are equal, then the result will be $setValue(getValue(a))$. Otherwise, the result will be $setValue(null)$.
$a = (a - b) \bigtriangledown (a \bigtriangleup b)$	If $getValue(a) = getValue(b)$, then $(a - b) \bigtriangledown (a \bigtriangleup b) = setValue(null) \bigtriangleup a = a$, else $(a - b) \bigtriangledown (a \bigtriangleup b) = a \bigtriangledown setValue(null) = a$.
$a - (a \bigtriangleup b) = a - b$	If $getValue(a) = getValue(b)$, then $a - (a \bigtriangleup b) = a - a = setValue(null)$ and $a - b = a - a = setValue(null)$, else $a - (a \bigtriangleup b) = a - setValue(null) = a$ and $a - b = a$.

- The 'set' type implementation (it is shown in table A.3). This implementation is oriented for elements whose *dataType* is a generic *set* of values. That is to say, this element may have multiple values. The generalization, subtraction and extension operators are implemented as the intersection,

Table A.3. The 'set' type implementation of the *metadataElement* ADT

Implementation of operations

Operation	Implementation
$emptyElement$	$setValue(\varnothing)$
$getValue(a)$	Obtain the 'set of values'.
$setValue(a, value)$	Set the 'set of values'.
$a \bigtriangledown b$	$setValue(getValue(a) \cup getValue(b))$ The result of union of two elements a and b is a new element whose values are the union of values of a and b. Example: $getValue(a) =< v1 >; getValue(b) =< v2 >; getValue(a \cup b) =< v1, v2 >$
$a \bigtriangleup b$	$setValue(getValue(a) \cap getValue(b))$ The result of the generalization of two metadata elements a and b is a new element whose values are the intersection of values of a and b. Example: $getValue(a) =< v1, v2 >$; $getValue(b) =< v1, v3 >$; $getValue(a \cap b) =< v1 >$
$a - b$	$setValue(getValue(a) - getValue(b))$ The result of the generalization of two metadata records a and b is a new element whose metadata elements values are the subtraction of values of a and b. Example: $getValue(a) =< v1, v2, v3 >$; $getValue(b) =< v3 >$; $getValue(a - b) =< v1, v2 >$

Proof of the equations for this implementation of operations

Equation	Proof
$a \bigtriangleup b = b \bigtriangleup a$	Derived from the implementation. The intersection of sets is commutative.
$(a \bigtriangleup b) \bigtriangleup c = a \bigtriangleup (b \bigtriangleup c)$	Derived from the implementation. The intersection of sets is associative.
$a = (a - b) \bigtriangledown (a \bigtriangleup b)$	Derived from the implementation. The intersection, subtraction and union of sets comply with this equivalence.
$a - (a \bigtriangleup b) = a - b$	Derived from the implementation. The intersection, subtraction and union of sets comply with this equivalence.

subtraction and union of the sets of values corresponding to the elements implied in the operation. An example of an element for which this implementation could apply is a *keywords* element (e.g., the *descriptivekeywords* attribute of the *MD_Identification* class in ISO 19115 (ISO, 2003a)) that contains a set of values describing the topics covered by the resource.

- The 'string' type implementation (it is shown in table A.4). This implementation for elements whose *dataType* is a *string*. It is assumed that this *dataType* has the following features: it can be assigned a *null* value; there is an operation for the concatenation of strings denoted as $\_ \circ \_$; the *commonPrefix(string, string)* operation returns a new *string* with the common prefix of the two *strings*; and the *removePrefix(string, string)* operation returns a new string removing from the first argument the beginning characters that correspond to the second argument (if it is really a prefix of the first argument). An example of an element for which this implementation could apply is the *title* element (e.g., the *title* attribute of the *CI_Citation* class in ISO 19115). Usually, a generic title is given for a collection and each component is entitled with the concatenation of the generic title plus the code of a specific dataset component (e.g., the numbering of a tile). Extension, subtraction and generalization operations are implemented in order to obtain these concatenations, suffixes and prefixes.

Table A.4. The 'string' type implementation of the *metadataElement* ADT

Implementation of operations

Operation	Implementation
emptyElement	*setValue(null)*
getValue(a)	Obtain the value.
setValue(a, value)	Set the value.
$a \bigtriangledown b$	*setValue(getValue(b) ○ getValue(a))* Example:*getValue(a)*= "A:28"; *getValue(b)*="National Topographic Map. "; *getValue(a $\bigtriangledown$ b)*="National Topographic Map. A:28"
$a \bigtriangleup b$	*setValue(commonPrefix(getValue(a), getValue(b)))* Example:*getValue(a)*="National Topographic Map. A:28"; *getValue(b)*="National Topographic Map. B:29"; *getValue(a $\bigtriangleup$ b)*="National Topographic Map. "
$a - b$	*setValue(removePrefix(getValue(a), commonPrefix(getValue(a) , getValue(b))))* Example:*getValue(a)*="National Topographic Map. A:28"; *getValue(b)*="National Topographic Map. ";*getValue(a − b)*="A:28"

Proof of the equations for this implementation of operations

Equation	Proof
$a \bigtriangleup b = b \bigtriangleup a$	$a \bigtriangleup b = setValue(commonPrefix(getValue(a), getValue(b))) =$ $setValue(commonPrefix(getValue(b), getValue(a))) = b \bigtriangleup a$
$(a \bigtriangleup b) \bigtriangleup c = a \bigtriangleup (b \bigtriangleup c)$	$(a \bigtriangleup b) \bigtriangleup c =$ $setValue(commonPrefix(getValue(a), getValue(b))) \bigtriangleup c =$ $setValue(commonPrefix(commonPrefix(getValue(a),$ $getValue(b)), getValue(c))) =$ $setValue(commonPrefix(getValue(a), getValue(b), getValue(c)));$ $a \bigtriangleup (b \bigtriangleup c) =$ $a \bigtriangleup setValue(commonPrefix(getValue(b), getValue(c))) =$ $setValue(commonPrefix(getValue(a), commonPrefix($ $getValue(b), getValue(c)))) =$ $setValue(commonPrefix(getValue(a), getValue(b), getValue(c)));$
$a = (a - b) \bigtriangledown (a \bigtriangleup b)$	$(a - b) \bigtriangledown (a \bigtriangleup b) =$ $setValue(removePrefix(getValue(a), commonPrefix(getValue(a)$ $, getValue(b))) \bigtriangledown setValue(commonPrefix(getValue(a)$ $, getValue(b))) =$ $setValue(commonPrefix(getValue(a), getValue(b))○$ $removePrefix(getValue(a), commonPrefix(getValue(a)$ $, getValue(b)))) =$ $setValue(getValue(a)) = a$ Example: *getValue(a)* ="title 1";*getValue(b)* ="title 2"; *getValue(a − b)* ="1";*getValue(a $\bigtriangleup$ b)* ="title "; *getValue((a − b) $\bigtriangledown$ (a $\bigtriangleup$ b))* ="title "○"1"="title 1"= *getValue(a)*
$a - (a \bigtriangleup b) = a - b$	$a - (a \bigtriangleup b) =$ $a - setValue(commonPrefix(getValue(a), getValue(b))) =$ $setValue(removePrefix(getValue(a), commonPrefix($ $getValue(a), commonPrefix(getValue(a), getValue(b))))) =$ $setValue(removePrefix(getValue(a), commonPrefix($ $getValue(a), getValue(b)))) =$ $a - b$

- The 'aggregation' type implementation (it is shown in table A.5). This implementation is oriented for *dataTypes* where an aggregated function can be defined. Examples of these aggregated functions are the maximum, minimum, sum, or average of numbers. But this is also applicable to other more sophisticated functions operating over 2D geometries or time intervals. This is particularly interesting for elements such as the *geographic location* of a resource, which is usually represented by means of a bounding box. In such a case, the *geographic location* of the collection can be com-

Table A.5. The 'aggregation' type implementation of the *metadataElement* ADT

Implementation of operations

Operation	Implementation
emptyElement	$setValue(null)$
getValue(a)	Obtain the value.
setValue(a, value)	Set the value.
$a \bigtriangledown b$	$setValue(getValue(a))$ It returns always the first operand.
$a \bigtriangleup b$	$setValue(aggFunction(getValue(a), getValue(b)))$
$a - b$	$setValue(getValue(a))$ It returns always the first operand.

Proof of the equations for this implementation of operations

$a \bigtriangleup b = b \bigtriangleup a$	The aggregated function is conmutative.
$(a \bigtriangleup b) \bigtriangleup c = a \bigtriangleup (b \bigtriangleup c)$	The aggregated function is associative.
$a = (a - b) \bigtriangledown (a \bigtriangleup b)$	$(a - b) \bigtriangledown (a \bigtriangleup b) = a \bigtriangledown (a \bigtriangleup b) = a$
$a - (a \bigtriangleup b) = a - b$	$a - (a \bigtriangleup b) = a$ $a - b = a$

puted as the envelope or minimum bounding box that covers the bounding
boxes of the components (see figure A.5). This case is also similar to the
temporal extent element describing the time interval for which a resource is
valid. The temporal extent of the collection is usually defined as the minimum time interval that covers the *temporal extent* of each component in
the collection. This aggregated function, denoted as *aggFunction* in table
A.5, is used as the implementation of the generalization.

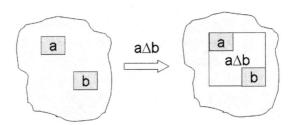

Fig. A.5. Generalization (envelope) of bounding boxes

As a conclusion, a system using implementations of the *metadataElement*
ADT could perform this reversible transformation. Given an original metadata
scenario (with all the metadata records), the user should decide what kind
of *metadataElement* implementation corresponds with each element of the
metadata standard used. And then, the system could minimize the original
metadata. Obviously, the user should select those implementations that are
more beneficial for minimization. For instance, if the user detects that all the
values of the element *abstract* are identical, he should opt for a 'default' type
implementation. In contrast, if he detects that the *keyword* element can have

multiple values and, apart from some exceptions, most of them are identical, he should opt for a 'set' type implementation.

This algebraic approach is interesting for such a reversible system. But being realistic, the minimization of an original scenario rarely takes place. That is precisely what metadata creators are trying to avoid. They want to create uniquely a minimized model that could be expanded to the complete metadata model in demand. Furthermore, the configuration of the system implying the selection of the appropriate implementation for each element should result really tedious. Thus, the Metadata Knowledge Base presented in chapter 2 is only focused in the reverse process, i.e. obtaining complete metadata from a minimized model and reducing the configuration tasks as much as possible. Notwithstanding that, this previous study with algebraic specifications of metadata records has contributed in important aspects of the knowledge base:

- The metadata stored in the knowledge base corresponds with a minimization scenario where the implementations allowed for each element e are:
  1. An implementation verifying that the generalization of the element e among
     $MD\_Collection$ and MD_is is is equals to the value of e in $MD\_Collection$:
     $getEl(commonMetadata, e) = getEl(MD\_Collection, e) \triangle getEl(MD\_1, e) \triangle \ldots \triangle$
     $getEl(MD\_n, e) = getEl(MD\_collection, e)$.
     Thus, the value of e within *common metadata* contains the original value of
     $MD\_Collection$. An example of this implementation is the selection of a 'string' type implementation for the *title* element whenever all the component titles have as prefix the collection title.
  2. Otherwise, a 'default' type implementation must be used. Thus, when all the records do not contain the same value for this element, the value contained within *common metadata* compartment will be null and the value in *collection-specific metadata* will contain the original value of $MD\_Collection$.
- There is a subgroup of the first allowed types of implementations that have interesting properties. This group of implementations is referred to as implementations producing *inherent metadata*. These implementations add two characteristics. Firstly, the subtraction operation always return the first operand ($a - b = a$). And secondly, the generalization of the element e in MD_is is is equals to the value of e in $MD\_Collection$:
  $getEl(MD\_1, e) \triangle \ldots \triangle getEl(MD\_n, e) = getEl(MD\_Collection, e) \triangle getEl(MD\_1, e) \triangle$
  $\ldots \triangle getEl(MD\_n, e) = getEl(MD\_collection, e)$.
  Looking back to figure A.2 we realize that these implementations do not minimize the values (the values of these elements in the left side or in the right side are identical). Thus, it is not necessary to store the original value of the element in the *common metadata* compartment. We can compute this value when needed. Finally, according to the possible existence of

these implementations, the *common metadata* in figure A.2 is divided into: *inherent metadata* containing the result of the generalization of this special subgroup of elements; and *coincident metadata* containing the rest of non-empty metadata.

- The metadata records describing the components correspond with the MDS_i depicted in figure A.2.
- The metadata record describing the collection ($MDS\_Collection$) contains the value for each element e as follows:
  - If it has a non-null value in *coincident metadata*, this value is stored in $MDS\_Collection$.
  - Otherwise, if the element has not been classified as producing *inherent metadata*, the value contained in $MDS\_Collection$ corresponds to the value in *collection-specific* metadata.

  That is to say, all the values of elements in $MDS\_Collection$ correspond to a part of the original values in $MD\_Collection$. The rest of original values are obtained by computing and adding the *inherent metadata* when needed. All the element values that are needed to restore the MDS_i (they were initially stored in *common metadata*) are in $MDS\_Collection$ too.
- The functions specified in the *_wholeInferredValuesSpecification* would correspond with the generalization operations of those elements classified as producing inherent metadata. We do not store the generalization of other elements because we are not going to minimize an original scenario, we are only interested in the reverse process and we already have the *coincident metadata*.
- The functions specified in the *_partDerivedValuesSpecification* attribute of $KB\_AggregationRelationType$ would correspond to the different implementations of the extension operation. The specification of the extension operation is not necessary for elements with a 'default' type implementation or for those elements classified as producing *inherent metadata*. In the first case, this extension operation is considered as a default mechanism: the inheritance by default of component records with respect to the collection record. That is to say, if a component has not got a value, it will try to obtain it from the collection record. In the second case, the components are also supposed to have the original value without further processing (the subtraction operation always returns the first operand for these implementations producing *inherent metadata*). This means that there can not be functions in both *_wholeInferredValuesSpecification* and *_partDerivedValuesSpecification* to obtain a value for the same element.
- We should assure that the functions specified in *_partDerivedValuesSpecification* and *_wholeInferredValuesSpecification* comply with the equations in the *metadataElement* ADT. For instance, the functions specified in *_wholeInferredValuesSpecification* must be commutative and associative. Additionally, although we do not specify all the operations (e.g., subtraction) it must be assured that we can find associated functions that comply with the specified equations.

- The process of establishing a mapping between an element and the appropriate implementation of the *metadataElement* ADT aids to identify the metadata inference that a concrete aggregation relation type should support.

A.2 Metadata Inference

A.2.1 Generation of complete values

KB_Metadata.getCompleteValues

```
/**
 * Returns the list of complete XML instances </p>
 * If _completeValues is not precalculated, it is generated again</p>
 * For the automatic generation, it uses the methods getWholeInferredValues and
 * getValuesBeingPart, which climb up and down through the whole part hierarchy
 * respectively.</p>
 */
public List getCompleteValues(){
  if (_completeValues == null)
  {
    // completeValues must be computed
    _completeValues = new LinkedList();
    // Obtain the values of the record acting as part
    List valuesBeingPart = getValuesBeingPart();
    // Obtain the values of the record acting as whole
    XML wholeInferredValues = null;
    if (getPartRelation()!=null)
      wholeInferredValues = getPartRelation().getWholeInferredValues();
    // Apply priorities
    // 2nd priority. The whole-inferred values have lower priority
    // 1st priority. The values obtained acting as part have higher priority
    ListIterator it = valuesBeingPart.listIterator();
    while(it.hasNext()) {
      XML completeXML = null;
      if (wholeInferredValues!=null){
        completeXML = wholeInferredValues.getCopy();
        completeXML.update( (XML) it.next());
      } else
        completeXML = (XML) it.next();
      _completeValues.add(completeXML);
    }
  }
  return _completeValues;
}
```

KB_Metadata.getValuesBeingWhole

```
/**
 * Returns the metadata of this record, acting this record as a collection
 * metadata record
 */
public XML getValuesBeingWhole() {
  // Obtain whole inferred values
  XML wholeInferredValues = null;
  if (getPartRelation()!=null)
    wholeInferredValues = getPartRelation().getWholeInferredValues();
  // apply priorities
  // 2nd priority. The whole-inferred values have lowest priority
  // 1st priority. The specific values have higher priority
  if (wholeInferredValues!=null) {
```

```
      wholeInferredValues.update(getSpecificValues());
      return wholeInferredValues;
   } else
      return getSpecificValues().getCopy();
}
```

KB_Metadata.getValuesBeingPart

```
/**
 * Returns the metadata of this record, acting this record as a part of
 * a collection
 * It takes into account that :a record may belong to several collections
 * and that a record may have several grandparents
 */
public List getValuesBeingPart(){
   List results = new LinkedList(); // initialization of list of results
   if ((getWholeRelations()==null)||(getWholeRelations().isEmpty()))
     results.add(getSpecificValues().getCopy());
   else {
    // a record may belong to several collections
    ListIterator it = getWholeRelations().listIterator();
    while (it.hasNext())
    {
      KB_AggregationRelation rel = (KB_AggregationRelation) it.next();
      // Obtain inherited XMLs (a record may have only one parent but several
      // grandparents)
      List partInheritedValues = rel.getPartInheritedValues();
      // Obtain derived XMLs (one for each possible inherited XML)
      List partDerivedValues = rel.getPartDerivedValues(getSpecificValues());
      // Apply priorities
      ListIterator itInh = partInheritedValues.listIterator();
      ListIterator itDer = partDerivedValues.listIterator();
      while (itInh.hasNext()&&itDer.hasNext())
      {
        // 3rd priority. Inherited values have the lowest priority
        XML result = (XML)itInh.next();
        // 2nd priority. Specific values have middle priority
        result.update(getSpecificValues());
        // 1st priority. Derived values have highest priority
        result.update( (XML) itDer.next());
        // add result to the list of results
        results.add(result);
      }
    }
   }
   return results;
}
```

KB_AggregationRelation.getWholeInferredValues

```
/**
 * It returns the whole inferred values.
 * If _wholeInferredValues is not precalculated, it is generated again</p>
 * It climbs down through the whole-part hierarchy
 */
public XML getWholeInferredValues() {
   if (_wholeInferredValues == null)
   {
     // _wholeInferredValues must be calculated
     if ((getParts()!=null)&&(!getParts().isEmpty())
         && (getWholeInferredValuesSpecification()!=null)
         && (!getWholeInferredValuesSpecification().isEmpty()))
     {
       List parts = getParts();
       // find the values to merge
       List valuesToMerge = new LinkedList();
```

```
            ListIterator it = parts.listIterator();
            while (it.hasNext())
               valuesToMerge.add( ((KB_Metadata) it.next()).getValuesBeingWhole());
            // apply wholeInferredValuesSpecification
            _wholeInferredValues = XML.inferWholeValues(valuesToMerge
                    ,getWholeInferredValuesSpecification());
        }
    }
    return _wholeInferredValues;
}
```

KB_AggregationRelation.getPartInheritedValues

```
/**
 * It returns the list of inherited XMLs
 * If _partInheritedValues is not precalculated, it is generated again</p>
 * It climbs up through the whole-part hierarchy
 */
public List getPartInheritedValues(){
  if (_partInheritedValues==null)
    // _partInheritedValues must be calculated
    _partInheritedValues = getWhole().getValuesBeingPart();
  // it returns a copy of each XML
  List result = new LinkedList();
  ListIterator it = _partInheritedValues.listIterator();
  while (it.hasNext())
    result.add( ((XML)it.next()).getCopy());
  return result;
}
```

KB_AggregationRelation.getPartDerivedValues

```
/**
 * It returns the part derived values
 * @param part specific values of the metadata record
 */
public List getPartDerivedValues(XML part) {
  List results = new LinkedList();
  List partInheritedValuesList = this.getPartInheritedValues();
  ListIterator it = partInheritedValuesList.listIterator();
  while ( it.hasNext())
  {
    XML partInheritedValues = (XML) it.next();
    results.add(XML.deriveValues(partInheritedValues,part
                       ,getPartDerivedValuesSpecification()));
  }
  return results;
}
```

A.2.2 Update of whole-part hierarchy

KB_Metadata.updateWholePartHierarchy

```
/** Update the whole-part hierarchy */
public void updateWholePartHierarchy() {
  updateWhole(); // update parents
  updateParts(); // update children
  updateCompleteValues(); // update complete values
}
```

KB_Metadata.updateWholes

```
/** Update parents */
public void updateWholes() {
  if ((getWholeRelations()!=null)&&(!getWholeRelations().isEmpty()))
  {
    // update the whole-inferred values for each relation where
    // 'this' is included
    ListIterator it = getWholeRelations().listIterator();
    ((KB_AggregationRelation)it.next()).updateWholeInferredValues();
  }
}
```

KB_Metadata.updateParts

```
/** Update parts */
public void updateParts() {
  if (getPartRelation()!=null)
    // update the partInheritedValues stored in the partRelation
    getPartRelation().updatePartInheritedValues();
}
```

KB_Metadata.updateCompleteValues

```
/** It provokes the recalculation of complete valures */
public void updateCompleteValues() {
  _completeValues = null;
  getCompleteValues();
}
```

KB_AggregationRelation.updateWholeInferredValues

```
/**
 * It implies the recalculation of _wholeInferredValues of this relation
 * and higher level relations
 * It is invoked from KB_Metadata.updateWhole when the specific values of a
 * metadata record have been updated, or this record has been added to a
 * collection.
 * It also implies the recalculation of completeValues in the ascending hierarchy
 */
public void updateWholeInferredValues() {
  if ((getWholeInferredValuesSpecification()!=null)&&
      (!getWholeInferredValuesSpecification().isEmpty()))
  {
    // prior value is invalidated
    _wholeInferredValues=null;
    // climb up through the whole-part hierarchy
    getWhole().updateWholes();
    // Then, climb down through the whole-part hierarchy to recalculate complete
    // values
    getWhole().updateCompleteValues();
  }
}
```

KB_AggregationRelation.updatePartInheritedValues

```
/**
 * It implies the recalculation of _partInherited values of this relation and lower
 * relations
 * It is invoked from KB_Metadata.updateParts when the specific values of a
 * metadata record have been updated, or this record has been added to a
 * collection
 */
public void updatePartInheritedValues() {
  // prior value is invalidated
  _partInheritedValues = null;
  // climb down through the whole-part hierarchy
```

```
    ListIterator it = getParts().listIterator();
    KB_Metadata part = null;
    while(it.hasNext()) {
        part = (KB_Metadata) it.next();
        // recursive invocation
        part.updateParts();
        // then, climb up updating the complete values
        part.updateCompleteValues();
    }
}
```

A.2.3 Example of a *wholeInferredValues* specification

```
<?xml version="1.0" encoding="ISO-8859-1"?>
<xsl:stylesheet version="1.0"
xmlns:xsl="http://www.w3.org/1999/XSL/Transform"
        xmlns:iso19115="http://www.isotc211.org/iso19115"
        xmlns:xsi="http://www.w3.org/2001/XMLSchema-instance"
exclude-result-prefixes="iso19115">

<xsl:output method="xml" indent="yes" encoding="ISO-8859-1"/>

<xsl:template match="/">
  <xsl:apply-templates select="components"/>
  <!-- components tag groups the individual metadata records, i.e. MD_Metadata
       tags -->
</xsl:template>

<xsl:template match="components">
 <!-- check whether the metadata of a component has geographic elements -->
 <xsl:if test="./MD_Metadata/identificationInfo/*/extent/geographicElement/*/
            northBoundLatitude">
  <!-- generate the tags of XML output -->
  <xsl:element name="iso19115:MD_Metadata">
   <xsl:element name="identificationInfo">
    <xsl:element name="iso19115:MD_DataIdentification">
     <xsl:element name="extent">
      <xsl:element name="geographicElement">
       <xsl:element name="iso19115:EX_GeographicBoundingBox">
        <xsl:variable name="total" select="count(./MD_Metadata/identificationInfo/
            */extent/geographicElement/EX_GeographicBoundingBox)"/>
        <!-- generate westBoundLongitude -->
        <xsl:element name="westBoundLongitude">
         <xsl:call-template name="agg">
          <xsl:with-param name="plist" select=
".MD_Metadata/identificationInfo/*/extent/geographicElement/EX_GeographicBoundingBox/
            westBoundLongitude"/>
          <xsl:with-param name="index" select="1"/>
          <xsl:with-param name="total" select="$total"/>
          <xsl:with-param name="aggFunction" select="'min'"/>
         </xsl:call-template>
        </xsl:element>
        <!-- generate eastBoundLongitude -->
        <xsl:element name="eastBoundLongitude">
         <xsl:call-template name="agg">
          <xsl:with-param name="plist" select=
".MD_Metadata/identificationInfo/*/extent/geographicElement/EX_GeographicBoundingBox/
            eastBoundLongitude"/>
          <xsl:with-param name="index" select="1"/>
          <xsl:with-param name="total" select="$total"/>
          <xsl:with-param name="aggFunction" select="'max'"/>
         </xsl:call-template>
        </xsl:element>
        <!-- generate southBoundLatitude -->
```

```
        <xsl:element name="southBoundLatitude">
        <xsl:call-template name="agg">
        <xsl:with-param name="plist" select=
"./MD_Metadata/identificationInfo/*/extent/geographicElement/EX_GeographicBoundingBox/
        southBoundLatitude"/>
        <xsl:with-param name="index" select="1"/>
        <xsl:with-param name="total" select="$total"/>
        <xsl:with-param name="aggFunction" select="'min'"/>
        </xsl:call-template>
        </xsl:element>
        <!-- generate northBoundLatitude -->
        <xsl:element name="northBoundLatitude">
        <xsl:call-template name="agg">
        <xsl:with-param name="plist" select=
"./MD_Metadata/identificationInfo/*/extent/geographicElement/EX_GeographicBoundingBox/
        northBoundLatitude"/>
        <xsl:with-param name="index" select="1"/>
        <xsl:with-param name="total" select="$total"/>
        <xsl:with-param name="aggFunction" select="'max'"/>
        </xsl:call-template>
        </xsl:element>
        <!-- generate end tags of XML output -->
        </xsl:element> <!-- EX_GeographicBoundingBox -->
      </xsl:element> <!-- geographicElement -->
     </xsl:element> <!-- extent -->
    </xsl:element>  <!-- iso19115:MD_DataIdentification -->
   </xsl:element>  <!-- identificationInfo-->
  </xsl:element> <!--/iso19115:MD_Metadata-->
 </xsl:if>
</xsl:template>

<!-- This template applies the aggregated functions, max or min,
over a list of parameter values --> <xsl:template name="agg">
  <xsl:param name="plist" select="/.." /> <!-- list of coordinate values -->
  <xsl:param name="index"/> <!-- index of element in the list that it is compared -->
  <xsl:param name="total"/> <!-- total number of elements in the list -->
  <xsl:param name="aggFunction"/> <!-- aggregated function -->
  <xsl:choose>
    <xsl:when test="$index=$total">
      <!-- base case -->
      <xsl:value-of select="$plist[$index]"/>
    </xsl:when>
    <xsl:otherwise>
      <!-- recursive step -->
      <xsl:variable name="aggValue">
        <xsl:call-template name="agg">
          <xsl:with-param name="plist" select="$plist"/>
          <xsl:with-param name="index" select="$index + 1"/>
          <xsl:with-param name="total" select="$total"/>
          <xsl:with-param name="aggFunction" select="$aggFunction"/>
        </xsl:call-template>
      </xsl:variable>
      <!-- select the aggregation function that must be applied -->
      <xsl:choose>
        <xsl:when test="$aggFunction = 'max'">
          <xsl:choose>
            <xsl:when test="$aggValue &gt; $plist[$index]">
              <xsl:value-of select="$aggValue"/>
            </xsl:when>
            <xsl:otherwise>
              <xsl:value-of select="$plist[$index]"/>
            </xsl:otherwise>
          </xsl:choose>
        </xsl:when>
        <xsl:when test="$aggFunction = 'min'">
          <xsl:choose>
            <xsl:when test="$aggValue &lt; $plist[$index]">
```

```
              <xsl:value-of select="$aggValue"/>
          </xsl:when>
          <xsl:otherwise>
            <xsl:value-of select="$plist[$index]"/>
          </xsl:otherwise>
        </xsl:choose>
      </xsl:when>
    </xsl:choose>
  </xsl:otherwise>
</xsl:choose>
</xsl:template>

</xsl:stylesheet>
```

B

Crosswalks

B.1 CSDGM→ISO 19115 stylesheet

```xml
<?xml version="1.0" encoding="ISO-8859-1"?>
<xsl:stylesheet version="1.0"
    xmlns:xsl="http://www.w3.org/1999/XSL/Transform"
    xmlns:rdf="http://www.w3.org/1999/02/22-rdf-syntax-ns#"
    xmlns:rdfs="http://www.w3.org/2000/01/rdf-schema#"
    xmlns:dc="http://purl.org/dc/elements/1.1"
    xmlns:dcterms="http://purl.org/dc/terms"
    xmlns:iso19115="http://www.isotc211.org/iso19115/"
    xmlns:xsi="http://www.w3.org/2001/XMLSchema-instance">
    <xsl:output method="xml" indent="yes" encoding="ISO-8859-1"/>
    <xsl:template match="/">
        <xsl:apply-templates select="metadata"/>
    </xsl:template>
    <!-- conversion of main METADATA section into MD_METADATA -->
    <xsl:template match="metadata">
        <xsl:element name="iso19115:MD_Metadata">
            <!--   ISO19115:_MD_IDENTIFICATION-->
            <xsl:element name="identificationInfo">
                <xsl:apply-templates select="idinfo"/>
            </xsl:element>
            ...
        </xsl:element>
    </xsl:template>
    <!-- conversion of IDINFO section into MD_IDENTIFICATION -->
    <xsl:template match="idinfo">
        <xsl:element name="iso19115:_MD_Identification">
            <xsl:attribute name="xsi:type">iso19115:MD_DataIdentification
            </xsl:attribute>
            <!--   conversion of CITATION subsection -->
            <xsl:element name="citation">
                <xsl:apply-templates select="citation/citeinfo"/>
            </xsl:element>
                        ...
        </xsl:element>
    </xsl:template>
    <!-- template for CITATION element -->
    <xsl:template match="citation/citeinfo | identAuth/citeinfo">
        <!-- TITLE -->
        <xsl:element name="title">
            <xsl:value-of select="title"/>
        </xsl:element>
        <!-- there is no ALTERNATETITLE in FGDC -->
```

```
<!-- conversion of DATE element (mandatory). When empty, it is generated
    by default -->
<xsl:element name="date">
    <xsl:element name="date">
        <xsl:choose>
            <xsl:when test="./pubdate!=''">
                <xsl:value-of select="pubdate"/>
            </xsl:when>
            <xsl:otherwise>0001-01-01</xsl:otherwise>
        </xsl:choose>
    </xsl:element>
    <xsl:element name="dateType">
        <xsl:text>publication</xsl:text>
    </xsl:element>
</xsl:element>
<!-- conversion of EDITION element -->
<xsl:if test="edition">
    <xsl:element name="edition">
        <xsl:value-of select="./edition"/>
    </xsl:element>
</xsl:if>
<!-- there is no EDITIONDATE element in FGDC -->
<!-- generation of IDENTIFIER element -->
<xsl:if test="citId">
    <xsl:element name="identifier">
        <xsl:element name="code">
            <xsl:value-of select="./citId"/>
        </xsl:element>
    </xsl:element>
</xsl:if>
<!-- conversion of ORIGINATOR into CITEDRESPONSIBLEPARTY element
    (role="originator") -->
<xsl:for-each select="origin">
    <xsl:if test="normalize-space(.)!=''">
        <xsl:element name="citedResponsibleParty">
            <xsl:element name="organisationName">
                <xsl:value-of select="."/>
            </xsl:element>
            <xsl:if test="/metadata/idinfo/citation/citeinfo[onlink]">
                <xsl:element name="contactInfo">
                    <xsl:element name="onlineResource">
                        <xsl:element name="linkage">
                            <xsl:value-of select="../onlink"/>
                        </xsl:element>
                    </xsl:element>
                </xsl:element>
            </xsl:if>
            <xsl:element name="role">
                <xsl:text>originator</xsl:text>
            </xsl:element>
        </xsl:element>
    </xsl:if>
</xsl:for-each>
<!-- conversion of PUBLISHER into CITEDRESPONSIBLEPARTY element
    (role="publisher") -->
<xsl:if test="/metadata/idinfo/citation/citeinfo[pubinfo]">
    <xsl:element name="citedResponsibleParty">
        <xsl:element name="organisationName">
            <xsl:value-of select="./pubinfo/publish"/>
        </xsl:element>
        <xsl:element name="contactInfo">
            <xsl:element name="address">
                <xsl:element name="city">
                    <xsl:value-of select="./pubinfo/pubplace"/>
                </xsl:element>
            </xsl:element>
        </xsl:element>
```

```
            <xsl:element name="role">
                <xsl:text>publisher</xsl:text>
            </xsl:element>
        </xsl:element>
    </xsl:if>
    <!-- conversion of GEOFORM into PRESENTATIONFORM -->
    <xsl:if test="geoform">
        <xsl:element name="presentationForm">
            <xsl:value-of select="./geoform"/>
        </xsl:element>
    </xsl:if>
    <!-- conversion of SERINFO into SERIES -->
    <xsl:if test="serinfo">
        <xsl:element name="series">
            <!-- the subelements of SERINFO are mandatory but optional in ISO
                CI_Series -->
            <xsl:element name="name">
                <xsl:value-of select="./serinfo/sername"/>
            </xsl:element>
            <xsl:element name="issueIdentification">
                <xsl:value-of select="./serinfo/issue"/>
            </xsl:element>
        </xsl:element>
    </xsl:if>
    <!-- conversion of OTHERCIT into OTHERCITATIONDETAILS-->
    <xsl:if test="othercit">
        <xsl:element name="otherCitationDetails">
            <xsl:value-of select="./othercit"/>
        </xsl:element>
    </xsl:if>
    <!-- conversion of ISBN -->
    <xsl:if test="isbn">
        <xsl:element name="ISBN">
            <xsl:value-of select="./isbn"/>
        </xsl:element>
    </xsl:if>
    <!-- conversion of ISSN -->
    <xsl:if test="issn">
        <xsl:element name="ISSN">
            <xsl:value-of select="./issn"/>
        </xsl:element>
    </xsl:if>
</xsl:template>
<!--...-->
</xsl:stylesheet>
```

B.2 ISO 19115→DC stylesheet

```
<?xml version="1.0" encoding="ISO-8850-1"?>
<xsl:stylesheet
    version="1.0" xmlns:xsl="http://www.w3.org/1999/XSL/Transform"
    xmlns:rdf="http://www.w3.org/1999/02/22-rdf-syntax-ns#"
    xmlns:rdfs="http://www.w3.org/2000/01/rdf-schema#"
    xmlns:dc="http://purl.org/dc/elements/1.1"
    xmlns:dcterms="http://purl.org/dc/terms"
    xmlns:iso19115="http://www.isotc211.org/iso19115/"
    xmlns:xsi="http://www.w3.org/2001/XMLSchema-instance">
  <xsl:output method="xml" indent="yes" encoding="ISO-8859-1"/>
  ...
  <xsl:template match="/">
    <xsl:apply-templates select="iso19115:MD_Metadata"/>
  </xsl:template>
  <xsl:template match="iso19115:MD_Metadata">
    <xsl:element name="rdf:RDF">
```

```
    <xsl:element name="rdf:Description">
  <!-- CONVERSION OF TITLE ELEMENT: For each occurrence of attribute
  title in CI_Citation entity, a DC:TITLE occurrence will be generated.-->
    <xsl:for-each select="./iso19115:_MD_Identification/citation/title">
      <xsl:element name="dc:title">
        <xsl:value-of select="normalize-space(.)"/>
      </xsl:element>
    </xsl:for-each>
  ...
  <!-- CONVERSION OF CREATOR ELEMENT: Each occurrence of MD_Metadata.identificationInfo
  .pointOfContact (CI_ResponsibleParty entity with role="originator") must be mapped
  to a single value of DC:CREATOR. If the CI_ResponsibleParty has been correctly
  completed, organisationName or individualName or positionName must contain a non-null
  value. The value of these attributes (in the order previously mentioned) will be
  used to create a DC:CREATOR element. -->
    <xsl:for-each select="./iso19115:_MD_Identification/pointOfContact">
      <xsl:if test="normalize-space(./role/CI_RoleCode_CodeList)='originator'">
        <xsl:element name="dc:creator">
          <xsl:choose>
            <xsl:when test="./organisationName">
              <xsl:value-of select="./organisationName"/>
            </xsl:when>
            <xsl:when test="./individualName">
              <xsl:value-of select="./individualName"/>
            </xsl:when>
            <xsl:when test="./positionName">
              <xsl:value-of select="./positionName"/>
            </xsl:when>
            <xsl:otherwise><!-- This should never happen. -->
            </xsl:otherwise>
          </xsl:choose>
        </xsl:element>
      </xsl:if>
    </xsl:for-each>
  ...
  </xsl:element>
  </xsl:element>
  </xsl:template>
</xsl:stylesheet>
```

B.3 DC→ISO 19115 stylesheet

```
<?xml version="1.0" encoding="ISO-8859-1"?>
<xsl:stylesheet
      version="1.0" xmlns:xsl="http://www.w3.org/1999/XSL/Transform"
      xmlns:rdf="http://www.w3.org/1999/02/22-rdf-syntax-ns#"
      xmlns:rdfs="http://www.w3.org/2000/01/rdf-schema#"
      xmlns:dc="http://purl.org/dc/elements/1.1"
      xmlns:dcterms="http://purl.org/dc/terms"
      xmlns:iso19115="http://www.isotc211.org/iso19115/"
      xmlns:xsi="http://www.w3.org/2001/XMLSchema-instance">
  <xsl:output indent="yes" encoding="ISO-8859-1"/>
  ...
  <xsl:template match="/">
    <xsl:apply-templates select="rdf:RDF"/>
  </xsl:template>
  <xsl:template match="rdf:RDF">
    <xsl:if test="rdf:Description">
    <xsl:element name="iso19115:MD_Metadata">
      <xsl:element name="iso19115:_MD_Identification">
      <xsl:attribute name="xsi:type">iso19115:MD_DataIdentificationType
      </xsl:attribute>
        <xsl:element name="citation">
      <!-- CONVERSION OF TITLE ELEMENT: The title attribute is mandatory within
```

```
        CI_Citation entity. If there is no value for DC:TITLE, "Default Title"
        will be generated. -->
         <xsl:choose>
           <xsl:when test="./rdf:Description/dc:title">
             <xsl:element name="title">
               <xsl:value-of select="./rdf:Description/dc:title"/>
             </xsl:element>
           </xsl:when>
           <xsl:otherwise>
             <xsl:text>Default Title</xsl:text>
           </xsl:otherwise>
         </xsl:choose>
         ...
       </xsl:element> <!-- citation -->
       ...
  <!-- CONVERSION OF CREATOR ELEMENT: This element is optional in both standards.
     For each occurrence of DC:CREATOR, a new pointOfContact will be created. The
     text of DC:CREATOR will correspond to the CI_ResponsibleParty.organisationName
     attribute. -->
       <xsl:for-each select="./rdf:Description/dc:creator">
         <xsl:element name="pointOfContact">
           <xsl:element name="role">
             <xsl:element name="CI_RoleCode_CodeList">
               <xsl:text>originator</xsl:text>
             </xsl:element>
           </xsl:element>
           <xsl:element name="organisationName">
             <xsl:value-of select="."/>
           </xsl:element>
         </xsl:element>
       </xsl:for-each>
       ...
     </xsl:element>  <!-- iso19115:_MD_Identification -->
   </xsl:element> <!-- iso19115:MD_Metadata -->
  ...
 </xsl:if> <!-- of: <xsl:if test="rdf:Description"-->
 </xsl:template>
  ...
</xsl:stylesheet>
```

C

Applications

C.1 Revision of geographic metadata editors

Given the increasing importance of geographic metadata, numerous software
packages (dedicated tools or plug-ins in GIS tools) have appeared during the
last decade for the creation of metadata. Due to the extended use of CSDGM
and the recency of ISO 19115, most of the metadata edition tools give only
support to the CSDGM standard. A detailed revision of CSDGM-based tools
can be found through the Web site of the FGDC [1]. However, nowadays most
of them tend to migrate to ISO 19115 as soon as possible.

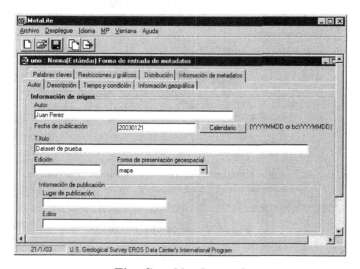

Fig. C.1. *MetaLite* tool

---

[1] Available at http://www.fgdc.gov/metadata/metatool.html.

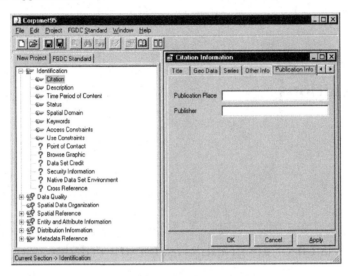

Fig. C.2. *CorpsMet95* tool

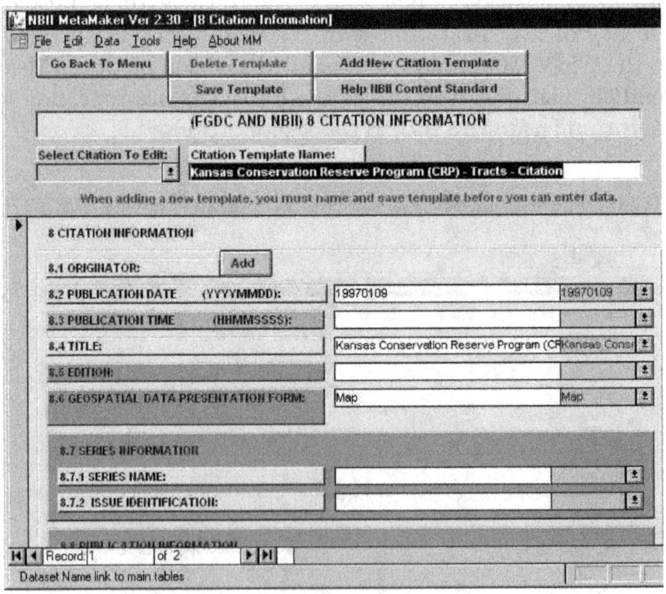

Fig. C.3. *MetaMaker* tool

Now, a reduced list of metadata edition tools will be briefly described. They have been selected by their relevance, extended use and their additional facilities for metadata creation. They are the following:

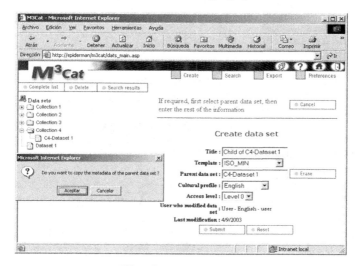

Fig. C.4. *M3Cat* tool

- One of the simplest but more extended tools is *MetaLite* [2] (see figure C.1), which has been developed by the FGDC and it is freely available. It only gives support for a minimum set of elements. It provides exchange in html, txt, sgml (xml) formats. This tool has been developed for Windows platforms (Visual basic) and stores metadata in an Access database. Additionally the application is delivered in 4 languages (es, en, fr, pt) and includes a small keywords dictionary in 4 languages.

- Another tool freely available is *Corpsmet95* (see figure C.2), which was developed by the U.S. Army Corps of Engineers [3]. It provides storage in text file (with extension *.met*) and works only in Windows platforms.

- *MetaMaker*[4] (see figure C.3) is also a freely available tool developed by the National Biological Information Infrastructure (NBII). It stores metadata in an Access database and can be operated in Windows platforms. Additionally it enables discovery of metadata records discovery by means of keyword searching.

- *MetaManager* [5] is another example that has been developed by a Cana dian company called Compusult. This tool also provides software to publish metadata records as a Clearinghouse node conforming to the Z39.50 search and retrieval protocol. This software acts as a bridge between spatial databases (ESRI SDE, ...) and a Clearinghouse gateway.

---

[2] http://edcnts11.cr.usgs.gov/MetaLite/
[3] http://www.usace.army.mil/
[4] http://www.umesc.usgs.gov/metamaker/nbiimker.html
[5] http://www.metadatamanager.com/

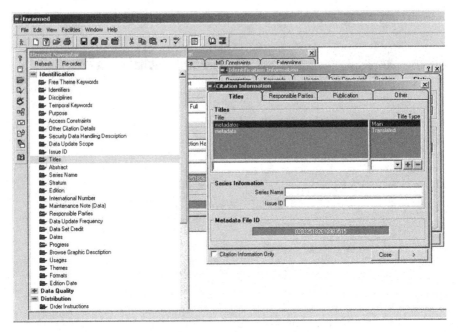

Fig. C.5. *Enraemed* tool

- *M3Cat* (see figure C.4) is a tool that has been developed a Canadian company called Intelec Geomatics Inc. [6]. This is a client-server Web application that stores metadata (according to different profiles and standards like ISO 19115 or CSDGM) in either an Access or an Oracle database. The most remarkable feature of this tool is that it gives support for hierarchical levels of metadata. However, at present this support only consists in a copy of metadata between parent and child datasets at the moment of child creation.
- *Enraemed* (see figure C.5) is a tool which was initially originated in 2000 by a project in Ethiopia, the Environmental Support Project carried out under Dutch-Ethiopian bi-lateral development cooperation. Later, the Global Spatial Data Infrastructure (GSDI), the FGDC and the United Nations Environment Programme agreed with Dutch-Ehtiopian governments for a technical exchange of this software. And nowadays, the software is being maintained and upgraded through the GSDI/FGDC. It is a client/server windows based application that supports ISO 19115 and CSDGM metadata. For metadata storage, SQL Server database is needed. Additionally, it gives support for metadata records discovery; it provides administration tools to create thesauri, and maps to help in the cataloguing process; and it is possible to configure different users of the application.

---

[6] http://www.intelec.ca/

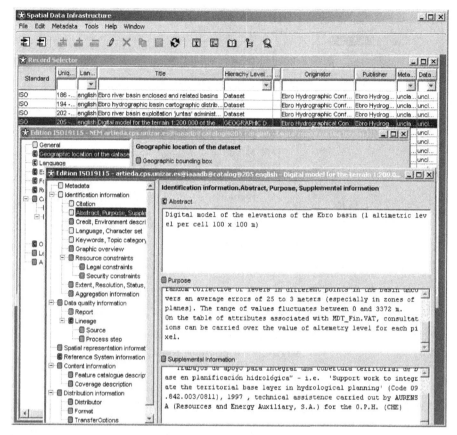

Fig. C.6. *CatMDEdit* tool

- *CatMDEdit* tool (see figure C.6) is an Open Source tool [7] which has been developed by TeiDE[8]. TeiDE is a Spanish consortium constituted by the R&D groups of the University of Zaragoza (Computer Science and Systems Engineering Dept.), the Jaume I University of Castellón (Dept. of Information Systems), and the Polytechnic University of Madrid (Topography and Cartography Dept.). CatMDEdit facilitates the documentation of resources, with special focus on the description of geographic information resources. Developed in Java, the main features of this tool are its multiplatform (Windows, Unix, Linux) and multilingual possibilities. Since the authors of this book, also members from TeiDE, have contributed to the development of this tool, further details of the tool can be found in section 5.3.

---

[7] http://catmdedit.sourceforge.net/

[8] http://teide.unizar.es/

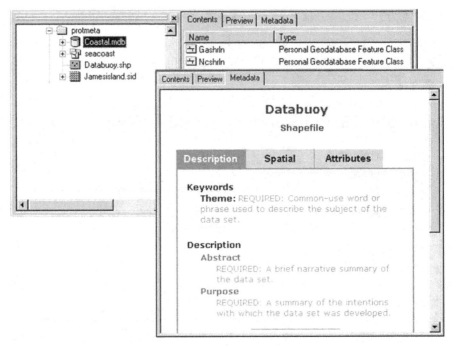

Fig. C.7. *ArcCatalog* tool

- And finally the *ArcCatalog* tool (see figure C.7), developed by ESRI[9], is perhaps one of the most widely used. Since the release of version 8.0 of Arc/Info, ArcCatalog enables metadata edition and automatic metadata generation for various types of sources (coverage, SDE, ...). It supports CSDGM (ESRI profile) and ISO 19115 (as much as possible). Metadata is stored usually in XML files together with dataset files, or inside the database for SDE One of the main features of ArcCatalog is the synchronization. Several metadata properties are automatically updated like the *spatial representation*, the *spatial reference system*, or the *entity and attribute information*. It also allows the creation of customized metadata editors (COM components) and presentation styles. And additionally the publishing of metadata records is possible by means of *ArcIMS*. The metadata generated by *ArcCatalog* can be integrated into *ArcIMS*.

C.2 Revision of thesaurus tools

The problem of creating appropriate content for thesauri has a great interest in the libraries field and other related disciplines. The fact to prove this interest

---

[9] http://www.esri.com

can be found in the increasing number of software packages that have appeared in last years for the construction of thesauri. For instance, the web site of Willpower Information [10] offers a detailed revision of more than 40 tools. Some tools are only available as a module of a complete information storage and retrieval system, but others also allow the possibility of working independently of any other software. And among these thesaurus creation tools, one may remark the following products:

- *BiblioTech*[11]. This is a multi-platform tool that forms part of BiblioTech PRO Integrated Library System and can be used to build an ANSI standard (Z39.19) thesaurus.
- *Lexico*[12]. This is a Java-based tool that can be accessed and/or manipulated over the Internet. It allows the definition of descriptive note fields that permit tracking of various details such as rationale for term selection, instructions for cataloging and retrieval, historical information, etc. This tool has been used by the U.S. Library of Congress to manage vocabularies and thesauri such as: the "Thesaurus for Graphic Materials", the "Global Legal Information Network Thesaurus", the "Legislative Indexing Vocabulary" and the "Symbols of American Libraries Listing".
- *MultiTes*[13]. This is a windows based tool that, among the main features, allows: support for an unlimited number of thesauri (both monolingual and multilingual); the automatic validation of conflicting relationships; up to 100 million terms per thesaurus and an unlimited number of hierarchies; and support for ANSI/NISO relationships plus user defined relationships and comment fields.
- *TermTree 2000*[14]. TermTree is a windows based tool that uses Access, SQL Server or Oracle for data storage. The tool verifies the validity of links as the thesaurus is created and automatically constructs all required reverse relationship links. Regarding import/export facilities, Term Tree 2000 can import/export TRIM thesauri[15] as well as a defined Term Tree 2000 tag format.
- *WebChoir*[16]. WebChoir is a family of client-server web applications that provide different utilities for thesaurus management. This family of tools supports multiple DBMS platforms. TermChoir is a hierarchical information organizing and searching tool that enables information professionals to create and search varieties of hierarchical subject categories, controlled vocabularies, and taxonomies based on either pre-defined standards or a

---

[10] http://www.willpower.demon.co.uk/thessoft.htm
[11] http://www.inmagic.com/
[12] http://www.pmei.com/lexico.html
[13] http://www.multites.com/
[14] http://www.termtree.com.au/
[15] Format used by the Towers Records Information Management system (http://www.towersoft.com/).
[16] http://www.webchoir.com

user-defined structure. LinkChoir is another tool that allows indexers to describe information sources using terminology organized in TermChoir. And SeekChoir is a retrieval system that enables users to browse thesaurus descriptors and their references (broader terms, related terms, synonyms, etc.), allowing the searcher many ways to investigate and employ related and synonymous topics and concepts while searching.

- *Synaptica* [17]. Synaptica is a client-server web application that can be installed locally on a client's intranet or extranet server. It has been developed with COM and Active Server Pages (ASP) technology and requires the installation of Internet Information Server (IIS), the web server of Microsoft. Regarding the storage, thesaurus data is stored in a SQL Server or Oracle database. The application supports the creation of electronic thesauri in compliance with ANSI/NISO Standard Z39.19-1993. The application also allows the exchange of thesauri in CSV (Comma-Separated Values) text format.

Another important aspect in thesaurus tools is the import/export capability. The main limitation with this respect is that the exchange format has not been standardized yet. The ISO norm for multilingual thesauri (ISO 5964) is currently undergoing review by ISO TC46/SC 9, and it is expected that among the new modifications it will include a standard exchange format for thesauri. It is believed that this format will be based on interoperable technologies like RDF/XML. In fact, some initiatives in this direction have already arisen:

- The ADL thesaurus Protocol (Janée et al., 2003) defines an XML and HTTP-based protocol for accessing thesauri. As a result of query operations, portions of the thesaurus are returned encoded in XML.
- The Language Independent Metadata Browsing of European Resources (LIMBER) project has published a Thesaurus Interchange Format in RDF (Matthews et al., 2002). Thesauri are used throughout the information retrieval world as a method of providing controlled vocabularies for indexing and querying. W3C is developing standards for the representation of ontologies to constrain the vocabularies of resource descriptions based on RDF. Such ontologies will allow distributed authoritative definition of vocabularies that support cross-referencing. And these ontology representations are planned to fulfill the role currently undertaken by thesauri. This work introduces an RDF representation of thesauri, which is proposed as a candidate thesaurus interchange format. This work also discusses whether it serves as a useful step on a migration path towards semantic web ontologies.
- The California Environmental Resources Evaluation System (CERES) and the NBII Biological Resources Division are collaborating in the

---

[17] http://www.synaptica.com/

CERES/NBII Thesaurus Partnership project [18] for the development of an Integrated Environmental Thesaurus and a Thesaurus Networking ToolSet for Metadata Development and Keyword Searching. One of the deliverables of this project is an RDF implementation of a representation of terms of a thesaurus.

- And finally, the "Semantic Web Activity: Advanced Development - Europe" (SWAD-Europe) project[19] is an EU-funded project (part of the IST-7 programme) also aiming at supporting the W3C's Semantic Web initiative in Europe. In particular, one of the activities of this project, SWAD-Europe Thesaurus Activity, has defined the Simple Knowledge Organization System (SKOS), a set of specifications and standards to support the use of knowledge organization systems (KOS) on the semantic web. And two of these specifications, SKOS Core and SKOS Mapping provide RDF vocabularies and models for describing thesaurus data.

Finally, it must be mentioned that, given that a thesaurus can be considered as an ontology specialized in organizing terminology (Gonzalo et al., 1998b), ontology editors could be another option for thesauri construction. (Denny, 2002) offers a detailed survey of ontology editors.

---

[18] http://ceres.ca.gov/thesaurus
[19] http://www.w3.org/2001/sw/Europe/

References

AENOR (1998). Mecanismo de Intercambio de Información Geográfica Relacional formado por Agregación (MIGRA), versión 1. UNE 148001 EXP 1998, Comité Técnico de Normalización 148 de AENOR(AEN/CTN 148).

Agirre E, Rigau G (1996). Word sense disambiguation using conceptual density. In *Proc. of the 16th International Conference on Computational Linguistics (Coling'96)*, pp 16–22, Copenhagen, Denmark.

Anaya D, Cantán O, Lacasta J, Nogueras J, Zarazaga FJ (2002). Interoperabilidad entre estándares de meta-datos geográficos. In *Proc. of the II Jornadas de Sistemas de Información Geográfica (JSIG'02)*, pp 73–86, El Escorial, Madrid.

Anderson WL (1997). Digital Libraries: a brief introduction. *SIGGROUP Bulletin*, 18(2).

ANSI (1993). Guidelines for the Construction, Format, and Management of Monolingual Thesauri. ANSI/NISO Z39.19-1993, American National Standards Institute. Revision of Z39.19-1980.

ANSI (1995). Information Retrieval Application Service Definition and Protocol Specification for Open Systems Interconnection. ANSI/NISO Z39.50-1995 (ISO 23950), American National Standards Institute. http://lcweb.loc.gov/z3950/agency/document.html.

ANSI (2001). The Dublin Core Metadata Element Set. ANSI/NISO Z39.85-2001, American National Standards Institute.

Arqued-Esquía VM, Zarazaga-Soria FJ, Losada-García JA (2001). El sistema de información GIS-Ebro. Metadatos y catálogo de datos geográficos. *BOLE.TIC*, 19:83–97.

Artale A, Franconi E, Guarino N, Pazzi L (1996). Part-Whole Relations in Object-Centered Systems: An Overview. *Data & Knowledge Engineering*, 20(3):347–383.

Baader F, Calvanese D, McGuinness D, Nardi D, Patel-Schneider P, (eds) (2003). *The Description Logic Handbook. Theory, Implementation and Applications*. Cambridge University Press.

Bañares JA, Bernabé MA, Gould M, Muro-Medrano PR, Nogueras FJ (2001). Infraestructura Nacional de Información Geográfica y su utilidad para las Administraciones Públicas. *BOLE.TIC*, 19:51–60.

Bañares JA, Zarazaga FJ, Nogueras J, Gutiérrez J, Muro-Medrano PR (2000). *Geographical Domain and Geographical Information Systems*, volume 19 of *GeoInfo Series*, chapter Construction and Use of Concept Hierarchies from Word Taxonomies for searching geospatial data, pp 5–8.

Backus JW, Bauer FL, Green J, Katz C, McCarthy J, Perlis AJ, Rutishauser H, Samelson K, Vauquois B, Wegstein JH, van Wijngaarden A, Woodger M, Nauer P (1963). Revised report on the algorithmic language Algol 60. *Comm. ACM*, 6(1):1–17.

Baeza-Yates R, Ribeiro-Neto B (1999). *Modern Information Retrieval*. New York. ACM Press, Addison Wesley.

Baker T, Dekkers M, Fischer T, Heery R (2003). Dublin Core Application Profile Guidelines. CWA 14855, CEN/ISSS Workshop - Metadata for Multimedia Information - Dublin Core.

Beaujardière J, (ed) (2002). *Web Map Server Implementation Specification. Version 1.1.1*. Open Geospatial Consortium Inc (Open GIS Consortium Inc), OpenGIS project document 01-068r3.

Bechhofer S, Goble C (2001). Thesaurus construction through knowledge representation. *Data & Knowledge Engineering*, 37(1):25–45.

Bechhofer S, van Harmelen F, Hendler J, Horrocks I, McGuinness DL, Patel-Schneider PF, Stein LA (2004). *OWL Web Ontology Language Reference*. W3C, W3C Recommendation 10 February 2004. http://www.w3.org/TR/2004/REC-owl-ref-20040210/.

Bernabé MA, Gould M, Granell C, Muro-Medrano PR, J. Nogueras FJZS (2002). A Spatial Data Catalogue Based Initiative to Launch the Spanish SDI. In *Proc. of the 6th Conference on Global Data Infrastructures (GDSI-6)*, Budapest, Hungary.

Bernabé MA, Gould M, Muro-Medrano PR, Nogueras J, Zarazaga FJ (2001). Effective steps toward the Spain National Geographic Information Infrastructure. In *Proc. of the 4th AGILE Conference on Geographic Information Science. GI in EUROPE: Integrative, Interoperable, Interactive*, pp 236–243, Brno, Czech Republic.

Berners-Lee T, Hendler J, Lassila O (2001). The Semantic Web. *Scientific American*.

Bertino E, Elmagarmid AK, Haeid MS (2001). Quality of Service in Multimedia Digital Libraries. *SIGMOD Record*, 30(1).

Béjar R, Aboal J, Gould M, Muro-Medrano PR, Vila P (2003a). Incremental Construction of a Regional SDI, an Example Case in the Galicia Region. In *Proc. of the 9th European Commission GI&GIS Workshop, ESDI: Serving the User*, A Coruña, Spain.

Béjar R, Gallardo P, Gould M, Muro-Medrano PR, Nogueras-Iso J, Zarazaga-Soria FJ (2004). A High Level Architecture for National SDI: the Spanish

Case. In *10th European Commission GI&GIS Workshop, ESDI: The State of the Art*, Warsaw, Poland.

Béjar R, Muro-Medrano PR, Nogueras-Iso J, Zarazaga-Soria FJ (2003b). Facilitating Uniform Intercommunity Natural Risk Management Through Metadata for High-Level Geographic Aggregations. In *Proc. of the 6th AGILE Conference on Geographic Information Science*, pp 663–672, Lyon, France.

Booch G, Rumbaugh J, Jacobson I (1998). *The Unified Modeling Language User Guide*. Addison-Wesley.

Booth D, Haas H, McCabe F, Newcomer E, Champion M, Ferris C, Orchard D, (eds) (2004). *Web Services Architecture*. W3C, W3C Working Group Note 11 February 2004. http://www.w3.org/TR/2004/NOTE-ws-arch-20040211/.

Borgida A, Brachman RJ (1993). Loading Data into Description Reasoners. In Buneman P, Jajodia S, (eds), *Proc. of the 1993 ACM SIGMOD International Conference on Management of Data*, pp 217–226, Washington, D.C. ACM Press.

Borgida A, Brachman RJ, McGuinness DL, Resnick LA (1989). CLASSIC: A Structural Data Model for Objects. In *Proceeedings of the 1989 ACM SIGMOD International Conference on Management of Data*, pp 59–67.

Borst P (1997). *Construction of Engineering Ontologies for Knowledge Sharing and Reuse*. PhD thesis, Twente University.

Boulos MNK, Roudsari AV, Carson ER (2001). Towards a Semantic Medical Web: HealthCyberMaps Dublin Core Ontology in Protégé-2000. In *Fifth International Protégé Workshop*, SCHIN, Newcastle, UK.

Box D, Ehnebuske D, Kakivaya G, Layman A, Mendelsohn N, Nielsen HF, Thatte S, Winer D (2000). Simple Object Access Protocol (SOAP) 1.1. W3C Note 8 May 2000, W3C. http://www.w3.org/TR/2000/NOTE-SOAP-20000508/.

Boxall J (2002). Geolibraries, the Global Spatial Data Infrastructure and Digital Earth: a time for map librarians to reflect upon the Moonshot. *International Journal of Special Libraries (INSPEL)*, 36(1):1–21.

Brachman RJ, Schmolze JG (1985). An Overview of the KL-ONE Knowledge Representation System. *Cognitive Science*, 9(2):171–216.

Bray T, Paoli J, Sperberg-McQueen CM, Maler E, (eds) (2000). *Extensible Markup Language (XML) 1.0 (Second Edition)*. W3C, W3C Recommendation 6 October 2000. http://www.w3.org/TR/2000/REC-xml-20001006.

Brickley D, Guha RV, (eds) (2004). *RDF Vocabulary Description Language 1.0: RDF Schema*. W3C, W3C Recommendation 10 February 2004. http://www.w3.org/TR/2004/REC-rdf-schema-20040210/.

Broekstra J, Kampman A, van Harmelen F (2002). Sesame: A Generic Architecture for Storing and Querying RDF and RDF Schema. In Horrocks I, Hendler J, (eds), *Proc. of the First Internation Semantic Web Conference*, number 2342 in Lecture Notes in Computer Science, pp 54–68. Springer Verlag.

Buehler K, McKee L, (eds) (1996). *The OpenGIS Guide. Introduction to Interoperable Geoprocessing. Part I of the Open Geodata Interoperability Specification (OGIS)*. OGIS Project 6 Technical Commitee of the OpenGIS Consortium Inc., OGIS TC Document 96-001.

Burstein M, Ankolenkar A, Paolucci M, Payne T, Sycara K, Lassila O, McIlraith S, Son TC, Zeng H, Hobbs J, Martin D, Narayanan S, McDermott D (2002). DAML-S: Semantic Markup for Web Services. Daml version 0.7, DAML.Org. http://www.daml.org/services/daml-s/0.7/daml-s.html.

Calvanese D, Giacomo GD, Lenzerini M, Vardi MY (2003). Reasoning on regular path queries. *SIGMOD Record*, 32(4):83–92.

Cantán O, Bañares JA, Gutierrez J, Nogueras J, Zarazaga FJ (2001a). Merging Catalog Services and GIS applications by component interoperability mechanisms. In *Proc. of the 7th European Commission GI&GIS Workshop, Managing the Mosaic*, Potsdam, Germany.

Cantán O, Casanovas M, Gutierrez J, Nogueras J, Zarazaga FJ (2001b). Joining Geographic Catalog Services and Map Servers with GIS applications. In *Proc. of the 4th AGILE Conference on Geographic Information Science. GI in EUROPE: Integrative, Interoperable, Interactive*, pp 347–354, Brno, Czech Republic.

Cantán O, Gutierrez J, López R, Nogueras J, Valiño J, Zarazaga FJ (2000). Servicios Distribuidos de Catálogo de Información Geográfica, una herramienta clave para el Conocimiento de la Información Territorial. In *Proc. of the II Conferencia Sobre Sistemas de Información Territorial, Territorial 2000.*, Pamplona, España.

Cantán O, Nogueras-Iso J, Torres MP, Zarazaga-Soria FJ, Lacasta J (2003). On the Problem of Finding the Geographic Data We Are Looking For. In *Proc. of the 9th European Commission GI&GIS Workshop, ESDI: Serving the User*, A Coruña, Spain.

CEN (1998). Geographic Information European Prestandards, Euro-norme Voluntaire for Geographic Information - Data description - Metadata. ENV 12657, European Committee for Standardization(CEN) - CEN/TC 287.

CEO (1999). A user guide provided by the center for earth observation programme (ceo programme) of the european commission. recommendations on metadata. describing the data, services and information you have available! (version 2.0). Center for Earth Observation(CEO).

Ceri S, Widom J (1993). Managing Semantic Heterogeneity with Production Rules and Persistent Queues. In *Proc. of the 19th Conference on Very Large Databases*, (Los Altos CA), Dublin. Morgan Kaufman.

Chandler A, Foley D, Hafez AM (2000). Mapping and Converting Essential Federal Geographic Data Committee (FGDC) Metadata into MARC21 and Dublin Core. Towards an Alternative to the FGDC Clearinghouse. *D-Lib Magazine*, 6(1).

Chapman RL, (ed) (1992). *Roget's International Thesaurus*. HarperCollins, 5th edition.

Chaudhri VK, Farquhar A, Fikes R, Karp PD, Rice JP (1998). Open Knowledge Base Connectivity 2.0. Technical Report KSL-98-06, Knowledge Systems Laboratory, Stanford, CA.

Christensen E, Curbera F, Meredith G, Weerawarana S (2001). *Web Services Description Language (WSDL) 1.1.* W3C, W3C Note 15 March 2001. http://www.w3.org/TR/2001/NOTE-wsdl-20010315.

Clark J, (ed) (1999). *XSL Transformations (XSLT) version 1.0.* W3C, W3C Recommendation 16 November 1999. http://www.w3.org/TR/1999/REC-xslt-19991116.

Clark J, DeRose S, (eds) (1999). *XML Path Language (XPath) version 1.0.* W3C, W3C Recommendation 16 November 1999. http://www.w3.org/TR/1999/REC-xpath-19991116.

Clark P, Thompson J, Holmback H, Duncan L (2000). Exploiting a thesaurus-based semantic net for knowledge-based search. In *Proc 12th Conf on Innovative Application of AI (AAAI/IAAI'00)*, pp 988–995.

Coleman DJ, Nebert DD (1998). Building a North American Spatial Data Infrastructure. *Cartography and Geographic Information Systems*, 25(3):151–160.

Common Logic Working Group (2004). Common Logic Standard. http://cl.tamu.edu/.

Cox S, Daisey P, Lake R, Portele C, Whiteside A, (eds) (2003). *OpenGIS Geography Markup Language (GML) Implementation Specification, Version 3.0.* Open Geospatial Consortium Inc (Open GIS Consortium Inc), OpenGIS Project Document OGC 02-023r4.

Craglia M (2001). Towards a European Approach to Metadata for Geographic Information. ETeMII project document, ETeMII. http://www.ec-gis.org:8080/wecgis/docs/F25765/D421-METADATA.PDF.

Craglia M, Annoni A, Masser I, (eds) (1999). *Geographic Information Policies in Europe: National and Regional Perspectives. EUROGI-EC Data Policy Workshop, Amersfoort 15th November 1999.* European Commission - Space Applications Institute, European Communities. http://www.ec-gis.org/reports/policies.pdf.

Croft WB (1995). NSF Center for Intelligent Information Retrieval. *Communications of the ACM*, 38(4).

Cuppens F, Demolombe R (1988). Cooperative answering: A methodology to provide intelligent access to databases. In *Proc. of the 2nd Int. Conf. Expert Database Systems*, pp 621–643, Firfax, VA.

DCMI (2001). Dublin Core Metadata Glossary, Final Draft. Dublin Core Metadata Inititiative (DCMI), http://library.csun.edu/mwoodley/dublincoreglossary.html.

DCMI (2004). Homepage of the Dublin Core Metadata Initiative. Dublin Core Metadata Initiative (DCMI), http://www.dublincore.org.

Decker S, Horrocks I, van Harmelen F, Decker S, Klein M (2000). OIL in a nutshell. In *Proc. of the 12th International Confererence on Knowledge Engineering and Knowledge Management*, Juan-les-Pins, France.

Denny M (2002). Ontology building: a Survey of Editing tools. *XML.com*. http://xml.com/pub/a/2002/11/06/ontologies.html.

EEA (2001). GEneral Multilingual Environmental Thesaurus (GEMET), version 2.0. European Environment Agency (EEA), European Topic Centre on Catalogue of Data Sources (ETC/CDS), http://www.mu.niedersachsen.de/cds/.

European Commission (1998). Public Sector Information: A Key Resource for Europe. Green Paper On Public Sector Information In The Information Society. COM 98/0585.

Fagin R, Kolaitis PG, Miller RJ, Popa L (2003). *Database Theory - ICDT 2003: 9th International Conference Siena, Italy, January 8-10, 2003. Proceedings*, volume 2572 / 2003, chapter Data Exchange: Semantics and Query Answering, pp 207 – 224. Springer-Verlag Heidelberg.

Farquhar A, Fikes R, Rice J (1996). The Ontolingua Server: A Tool for Collaborative Ontology Construction. Technical Report KSL 96-26, Stanford University, Knowledge Systems Laboratory.

Fellbaum C, (ed) (1998). *WordNet. An Electronic Lexical Database*. MIT Press.

Fernández P, Béjar R, Latre MA, Valiño J, Bañares JA, Muro-Medrano PR (2000). Web mapping interoperability in practice, a Java approach guided by the OpenGis Web Map Server Interface Specification. In *Proc. of the 6th European Commission GI&GIS Workshop, The Spatial Information Society - Shaping the Future*, Lyon, France.

FGDC (1998). Content Standard for Digital Geospatial Metadata, version 2.0. Document FGDC-STD-001-1998, Federal Geographic Data Committee, Metadata Ad Hoc Working Group.

FGDC (2000). *Content Standard for Digital Geospatial Metadata Workbook, version 2.0*. Federal Geographic Data Committee, Metadata Ad Hoc Working Group.

FGDC (2002). Content Standard for Digital Geospatial Metadata: Extensions for Remote Sensing Metadata. Public review draft, Federal Geographic Data Committee, Standards Working Group.

Fielding R, Gettys J, Mogul J, Frystyk H, Masinter L, Leach P, Berners-Lee T (1999). Hypertext Transfer Protocol - HTTP/1.1. RFC 2616, Internet Engineering Task Force (IETF). http://www.ietf.org/rfc/rfc2616.txt.

Fikes R, Kehler T (1985). The role of frame based representation in reasoning. *Communications of ACM*, 28(9):904–920.

Fonseca FT (2001). *Ontology-Driven Geographic Information Systems*. PhD thesis, The University of Maine, Orono, Maine.

Fonseca FT, Egenhofer MJ, Davis Jr. CA, Borges KAV (2000). Ontologies and knowledge sharing in urban GIS. *Computers, Environment and Urban Systems*, 24:251–271.

Forbus KD, de Kleer J (1993). *Building Problem Solvers*. MIT Press.

Frawley WJ, Piatetsky-Shapiro G, Matheus CJ (1991). *Knowledge Discovery in Databases*, chapter Knowledge Discovery in Databases: An Overview, pp 1–30. AAAI/MIT Press.

Fredikson A, North C, Schneiderman CPB (1999). Temporal, Geographical and Categorical Aggregations Viewed through Coordinated Displays: a case study with Highway Incident Data. In *Proc. of the 1999 Workshop on New Paradigms in Information Visualization and Manipulation in conjunction with the 8th ACM International Conference on Information and Knowledge Management*, pp 26–34, Kansas City, Missouri, United States.

Friedman-Hill E (2003). *Jess in Action: Rule-Based Systems in Java*. Manning Publication Co.

Gale W, Church KW, Yarowsky D (1992). A method for disambiguating word senses in a large corpus. *Computers and the Humanities*, 26:415–439.

Gardels K (1997). Open GIS and on-line environmental libraries. *ACM SIGMOD*, 26(1).

Geisler G, Giersch S, McArthur D, McClelland M (2002). Creating Virtual Collections in Digital Libraries: Benefits and Implementation Issues. In *Proc. of the second ACM/IEEE-CS joint conference on Digital libraries*, pp 210–218, Portland, Oregon, USA. ACM Press.

Genesereth M, Nilsson N (1987). *Logical Foundations of Artificial Intelligence*. Morgan Kaufmann.

Genesereth MR, Fikes RE (1992). Knowledge Interchange Format, Version 3.0 Reference Manual. Technical Report Logic-92-1, Computer Science Department, Stanford University.

GeoConnections (2001). *The Access Technical Services Manual Version 1.1*. http://www.geoconnections.ca/.

Giarratano J, Riley G (1998). *Expert Systems: Principles and Programming*. PWS-Kent, Boston, MA., 3rd edition.

Gonzalo J, Verdejo F, Chugur I, Cigarran J (1998a). Indexing with WordNet synsets can improve Text Retrieval. In *Proc. COLING/ACL'98 Workshop on Usage of WordNet for Natural Language Processing*.

Gonzalo J, Verdejo F, Peters C, Calzolari N (1998b). Applying EuroWordNet to Cross-Language Text Retrieval. *Computers and the Humanities*, Special Issue on EuroWord-Net.

Goodchild MF (1998). *Innovations in GIS 5*, chapter The Geolibrary, pp 59–68. Taylor and Francis, London.

Goodchild MF, Zhou J (2003). Finding Geographic Information: Collection-Level Metadata. *GeoInformatica*, 7(2):95–112.

Gordon AI (2000). *The COM and COM+ Programming Primer*. Prentice Hall PTR, 1st edition.

Gould M, Bernabé MA, Granell C, Muro-Medrano PR, Nogueras J, Rebollo C, Zarazaga FJ (2002). Reverse engineering SDI: Standards based Components for Prototyping. In *Proc. of the 8th European Comission GI&GIS Workshop, ESDI - A Work in Progress*, Dublin, Ireland.

Graham   P   (1998).   Digital   Strategies   for   the   Rutgers
   University   Libraries.   White   paper,   Rutgers   University.
   http://web.syr.edu/~psgraham/pgsite/pglibwork/pgrutgers/rul/digstratrul.pdf.

Graham S, Simeonov S, Boubez T, Daniels G, Davis D, Nakamura Y, Neyama
   R (2002). *Building Web Services with Java: Making Sense of XML, SOAP,
   WSDL and UDDI.* SAMS.

Gravano L, Chang CCK, García-Molina H, Paepcke A (1997). STARTS:
   Stanford Proposal for Internet Meta-Searching. In *Proc. of the 1997 ACM
   SIGMOD Conference.*

Groot R, McLaughlin J (2000). *Geospatial Data Infrastructure: concepts, cases
   and good practice.* Oxford University Press, New York, USA.

Gruber T (1992). A translation approach to portable ontology specifications.
   Technical Report KSL 92-71, Knowledge Systems Laboratory, Standford
   University, Stanford, CA.

Guarino N, Giaretta P (1995). *Towards Very Large Knowledge Bases*, chapter
   Ontologies and Knowledge bases: Towards a Terminological Clarification.
   IOS Press.

Guttag J, Horning JJ (1978). The algebraic specification of abstract data
   types. *Acta Informatica*, 10:27–52.

Guttag J, Horning JJ (1980). Formal specifications as a design tool. In *Proc.
   7h Syrup. On Principles of Programming (POPL)*, Las Vegas.

Haas H, Brown A, (eds) (2004). *Web Services Glossary.* W3C, W3C Working
   Group Note 11 February 2004. http://www.w3.org/TR/2004/NOTE-ws-
   gloss-20040211/.

Han J, Huang Y, Cercone N, Fu Y (1996). Intelligent Query Answering
   by Knowledge Discovery Techniques. *IEEE Trans. Knowl. Data Eng.*,
   8(3):373–390.

Heflin J, Hendler J (2000). Semantic Interoperability on the Web. In Association GC, (ed), *Proc. of Extreme Markup Languages 2000*, pp 111–120,
   Alexandria, VA.

Hill   L   (2002).   Alexandria   Digital   Library   Feature   Type
   Thesaurus.   Version   of   July   3,   2002,   University   of   California   at   Santa   Barbara,   Alexandria   Digital   Library.
   http://www.alexandria.ucsb.edu/gazetteer/FeatureTypes/ver070302/index.htm.

Hill LL, Janée G, Dolin R, Frew J, Larsgaard M (1999). Collection Metadata
   Solutions for Digital Library Applications. *Journal of the American society
   for Information Science*, 50(13):1169–1181.

Horebeek I, Lewi J (1989). *Algebraic Specifications in Software Engineering.*
   Springer, Berlin, Germany.

Hunter J (2001). MetaNet - A Metadata Term Thesaurus to Enable Semantic
   Interoperability Between Metadata Domains. *Journal of Digital information*, 1(8).

IEEE (1990). *IEEE Standard Computer Dictionary: A Compilation of IEEE
   Standard Computer Glossaries.* Institute of Electrical and Electronics Engineers, New York, NY.

ISO (1985). Documentation: Guidelines for the establishment and development of multilingual thesauri. ISO 5964, International Organization for Standardization.

ISO (1986). Documentation: Guidelines for the establishment and development of monolingual thesauri. ISO 2788, International Organization for Standardization.

ISO (1990). Information technology - Open Systems Interconnection - Specification of Basic Encoding Rules for Abstract Syntax Notation One (ASN.1). ISO/IEC 8825:1990, International Organization for Standardization.

ISO (2002). Geographic information - Services. ISO/DIS 19119, International Organization for Standardization, ISO/TC 211.

ISO (2003a). Geographic information - Metadata. ISO 19115:2003, International Organization for Standardization.

ISO (2003b). Geographic information - Metadata - Implementation specification. ISO/WD 19139, International Organization for Standardization, ISO/TC 211.

ISO (2003c). Geographic information - Spatial schema. ISO 19107:2003, International Organization for Standardization.

ISO (2003d). Information and documentation - The Dublin Core metadata element set. ISO 15836:2003, International Organization for Standardization.

ISO (2003e). Information technology - Metadata registries (MDR) - Part 3: Registry metamodel and basic attributes. ISO/IEC 11179-3:2003, International Organization for Standardization.

Janée G, Frew J (2002). The ADEPT digital library architecture. In *Proc. of the second ACM/IEEE-CS joint conference on Digital libraries*, pp 342 – 350, Portland, Oregon, USA.

Janée G, Ikeda S, Hill LL (2003). The ADL Thesaurus Protocol. http://www.alexandria.ucsb.edu/~gjanee/thesaurus/specification.html.

Keene SE (1989). *Object-Oriented Programming in Common Lisp. A Programmer's Guide to CLOS*. Addison-Wesley.

Kifer M, Lausen G, Wu J (1995). Logical Foundations of Object-Oriented and Frame-Based Languages. *Journal of the Association for Computing Machinery*.

Kolodziej K, (ed) (2003). *Open GIS Web Map Server Cookbook. Version: 1.0.1. Stage: Draft*. Open Geospatial Consortium Inc (Open GIS Consortium Inc), OGC Document Number 03-050r1.

Kottman C, (ed) (1999). *The OpenGIS Abstract Specification. Topic13: Catalog Services (version 4)*. Open Geospatial Consortium Inc (Open GIS Consortium Inc), OpenGIS Project Document 99-113.

Krovetz R, Croft WB (1992). Lexical amibuity in information retrieval. *ACM Transactions on Information Systems*, 10(2):115–141.

Lacasta J, Nogueras-Iso J, Torres MP, Zarazaga-Soria FJ (2003). Towards the geographic metadata standard interoperability. In *Proc. of the 6th AGILE Conference on Geographic Information Science*, pp 555–565, Lyon, France.

Lagoze C, Fielding D (1998). Defining Collections in Distributed Digital Ge-olibraries. *D-Lib Magazine*, 4(11).

Lesk M (1986). Automatic sense disambiguation Using Machine Readable Dictionaries:: How to tell a pine cone from an ice cream cone. In *Proc. of the 1986 SIGDOC Conference*, pp 24–26, New York. ACM Press.

Lieberman J (2003). OpenGIS Web Services Architecture. OpenGIS Discussion Paper OGC 03-025, Open Geospatial Consortium Inc (Open GIS Consortium Inc).

Longley PA, Goodchild MF, Maguire DJ, Rhind DW (2001). *Geographic Information Systems and Science. Chapter 7.*

LONGMAN (2003). *Longman Dictionary of Contemporary English.* Pearson ESL.

Mahesh K, Kud J, Dixon P (1999). Oracle at TREC8: a lexical approach. In *Proc. of 8th Text REtrieval Conference (TREC-8)*, pp 207–216, Maryland.

Manola F, Miller E, (eds) (2004). *RDF Primer.* W3C, W3C Recommendation 10 February 2004. http://www.w3.org/TR/2004/REC-rdf-primer-20040210/.

Marshall CC (1998). Making Metadata: A Study of Metadata Creation for a Mixed Physical-Digital Collection. In *Proc. of the ACM Digital Libraries '98 Conference*, pp 162–171, Pittsburgh, PA.

Mata EJ, Ansó J, Bañares JA, Muro-Medrano PR, Rubio J (2001). Enriquecimiento de tesauros con WordNet: una aproximación heurística. In *Proc. of the IX Conferencia de la Asociación Española para la Inteligencia Artificial (CAEPIA)*, pp 593–602, Gijón, Spain.

Matthews BM, Miller K, Wilson MD (2002). A Thesaurus Interchange Format in RDF. http://www.limber.rl.ac.uk/External/SW_conf_thes_paper.htm.

May W (2004). XPath-Logic and XPathLog: a logic-programming-style XML data manipulation language. *Theory and Practice of Logic Programming (TPLP)*, 4(3):239–287.

Mena E (1998). *OBSERVER: An Approach for Query Processing in Global Information Systems based on Interoperation across Pre-existing Ontologies.* PhD thesis, University of Zaragoza.

Miller GA (1990). WordNet: An on-line lexical database. *Int. J. Lexicography*, 3.

Miller GA, Leacok C, Tengi R, Bunker R (1993). A semantic concordance. In *Proc. of the ARPA Workshop on Human Language Technology.* Morgan Kauffman.

Minsky M (1981). *Mind design*, chapter A framework for representing knowledge, pp 95–128. MIT Press, Cambridge MA.

Méndez-Rodriguez EM (2002). *Metadatos y recuperación de información. Estándares, problemas y aplicabilidad en bibliotecas digitales*, chapter Normalización, pp 189–226. Ediciones Trea, S.L.

Murata M, Lee D, Mani M (2001). Taxonomy of XML Schema Languages using Formal Language Theory. In *Extreme Markup Languages*, Montreal, Canada.

Muro-Medrano PR, Nogueras-Iso J, Torres MP, Zarazaga-Soria FJ (2003). Web Catalog Services Of Geographic Information, An Opengis Based Approach In Benefit Of Interoperability. In *Proc. of the 6th AGILE Conference on Geographic Information Science*, pp 169–177, Lyon, France.

Nassar N (1997). Searching With Isearch, Moving beyond WAIS. *Web Techniques magazine, www.webtechniques.com.*

Nebert D, (ed) (2001). *Developing Spatial Data Infrastructures: The SDI Cookbook v.1.1.* Global Spatial Data Infrastructure. http://www.gsdi.org.

Nebert D, (ed) (2002). *OpenGIS Catalog Services Specification, Version 1.1.1.* Open Geospatial Consortium Inc (Open GIS Consortium Inc), OpenGIS project document 02-087r3.

Nebert D, Whiteside A, (eds) (2004). *OpenGIS - Catalogue Services Specification (version: 2.0).* Open Geospatial Consortium Inc (Open GIS Consortium Inc), OpenGIS Project Document 04-021r2.

Nogueras J, Cantán O, Bobadilla JM, Casanovas M, Zarazaga FJ (2000). CatArcConnector, un componente para el acceso a servicios de catálogo distribuido compatibles con la OpenGIS Catalog Interface Specification desde ArcView y ArcInfo. In *Proc. of the IX Conferencia Nacional de ESRI*, Madrid, España.

Nogueras J, Latre MA, Navas M, Rioja R, Muro-Medrano PR (2001a). Towards the construction of the Spanish National Geographic Information Infrastructure. In *Proc. of the 7th European Commission GI & GIS Workshop, Managing the Mosaic*, Potsdam, Germany.

Nogueras J, López R, O.Cantan, Zarazaga FJ, Gutierrez J (2001b). *Sistemas de Información Geográfica: Una aproximación desde la Ingeniería del Software y las Bases de Datos*, volume 2 of *Monografías y Publicaciones. Colección Ingeniería Informática.*, chapter Servicios de Catálogo de Información Geográfica y sus Infraestructuras de Apoyo, un Perfil Java para la Especificación Coarse-Grain de OpenGIS, pp 63–77. Madrid, España.

Nogueras-Iso J, Bañares JA, Lacasta J, Zarazaga-Soria J (2003a). A software tool for thesauri management, browsing and supporting advanced searches. In *Proc. of the GI-days 2003 conference*, pp 105–118, Münster, Germany.

Nogueras-Iso J, Lacasta J, Bañares JA, Muro-Medrano PR, Zarazaga-Soria FJ (2003b). Exploiting disambiguated thesauri for information retrieval in metadata catalogs. In *Proc. of the X Conferencia de la Asociación Española para la Inteligencia Artificial (CAEPIA)*, pp 279–290, San Sebastian, Spain.

Nogueras-Iso J, Lacasta J, Bañares JA, Muro-Medrano PR, Zarazaga-Soria FJ (2004a). Exploiting disambiguated thesauri for information retrieval in metadata catalogs. *Lecture Notes on Artificial Intelligence (LNAI)*, 3040:322–333.

Nogueras-Iso J, Latre MA, Muro-Medrano PR, Zarazaga-Soria FJ (2004b). Building eGovernment services over Spatial Data Infrastructures. In Traunmüller R, (ed), *3rd International Conference on Electronic Government (EGOV'04)*, volume 3183 of *Lecture Notes in Computer Science*, pp 387–391, Zaragoza, Spain.

Nogueras-Iso J, Zarazaga-Soria FJ, Béjar R, Álvarez P. R. Muro-Medrano PJ (2004c). OGC Catalog Services: a key element for the development of Spatial Data Infrastructures. *Computers and Geosciences.*

Nogueras-Iso J, Zarazaga-Soria FJ, Lacasta J, Béjar R, Muro-Medrano PR (2004d). Metadata Standard Interoperability: Application in the Geographic Information Domain. *Computers, Environment and Urban Systems*, 28(6):611–634.

Nogueras-Iso J, Zarazaga-Soria FJ, Lacasta J, Tolosana R, Muro-Medrano PR (2004e). Improving multilingual catalog search services by means of multilingual thesaurus disambiguation. In *10th European Commission GI&GIS Workshop, ESDI: The State of the Art*, Warsaw, Poland.

Nogueras-Iso J, Zarazaga-Soria FJ, Muro-Medrano P (2004f). Management of nested collections of resources in Spatial Data Infrastructures. In *1st International Workshop on Geographic Information Management (GIM'04)*, pp 878–882, Zaragoza, Spain. IEEE Computer Society.

Noy NF, Fergerson RW, Musen MA (2000). The knowledge model of Protege-2000: Combining interoperability and flexibility. In *2th International Conference on Knowledge Engineering and Knowledge Management (EKAW'2000)*, Juan-les-Pins, France.

OASIS (2004). Universal Discovery, Description and Integration of Web Services (UDDI) protocol. http://www.uddi.org/.

Official Journal of the European Union (2003). Directive 2003/98/EC of the European Parliament and of the Council of 17 November 2003 on the re-use of public sector information. L 345 , 31/12/2003 pp. 0090-0096.

Orfali R, Harkey D, Edwards J (1999). *Client/Server Survival Guide.* John Wiley & Sons, 3rd edition.

Ostman A, Nogueras J, Winter S (2002). Barriers for the implementation of GI standards and interoperability. In *Proc. of the 5th AGILE Conference on Geographic Information Science*, pp 307–313, Palma de Mallorca, España.

Pierre MS, LaPlant W (1998). Issues in crosswalking Content Metadata Standards. Niso standards white papers, National Information Standards Organisation. http://www.niso.org/press/whitepapers/crsswalk.html.

Popa L, Velegrakis Y, Hernandez M, Miller RJ, Fagin R (2002). Translating Web Data. In *28th International Conference for Very Large Databases (VLDB 2002).*

Powell A, Heaney M, Dempsey L (2000). RSLP Collection Description. *D-Lib Magazine*, 6(9).

Pundt H, Bishr Y (2002). Domain ontologies for data sharing. An example from environmental monitoring using field GIS. *Computer & Geosciences.*

Quillian MR (1967). Word Concepts: A Theory and Simulation of some Basic Semantic Capabilities. *Behavioral Science*, 12:410–430.

Rada R, Mili H, Bicknell E, Blettner M (1989). Development and application of a metric on semantic nets. *IEEE Transactions on Systems, Man and Cybernetic*, 19(1):17–30.

Radford I, Arranz VM, Ananiadou S, Tsujii J (1995). *Analisi Statistica dei Dati Testuali*, volume 1, chapter Dynamic context matching for knowledge acquisition from small corpora. CISU, Rome, Italy.

Rajabifard A, Williamson IP, Holland P, Johnstone G (2000). From Local to Global SDI initiatives: a pyramid of building blocks. In *Proc. of the 4th GSDI Conference*, Cape Town, South Africa.

Ramamohanarao K, Harland J (1994). An Introduction to Deductive Database Languages and Systems. *VLDB Journal*, 3:107–122.

Reich L, Vretanos PA, (eds) (2001). *OGC Web Services Stateless Catalog Profile (was Web Registry Service), Version 0.06*. Open Geospatial Consortium Inc (Open GIS Consortium Inc), OGC-IP Draft Candidate Specification OGC 01-062.

Resnik P (1995a). Disambiguating noun groupings with respect to WordNet senses. In *Proc. of the 3rd Workshop on Very Large Corpora*. MIT.

Resnik P (1995b). Using information content to evaluate semantic similarity in a taxonomy. In *Proc. of the 14th International Joint Conference on Artificial Intelligence (IJCAI)*, Montreal, August.

Resnik P, Yarowsky D (1997). A perspective on word sense disambiguation methods and their evaluation. In Light M, (ed), *ACL SIGLEX Workshop on Tagging Text with Lexical Semantics: Why, What and How?*, pp 79–86, Washington, D.C.

Rigaux P, Scholl M, Voisard A (2002). *Spatial Databases with application to GIS*. Morgan Kaufmann.

Salton G, (ed) (1971). *The SMART retrieval system - Experiments in Automatic Document Processing*. Prentice Hall, Inc., Englewood Cliffs, NJ.

Salton G, Buckley C (1988). Term-weighting approaches in automatic text retrieval. *Information Processing & Management*, 24:513–523.

Salton G, Lesk ME (1968). Computer evaluation of indexing and text precising. *Journal of the Association of Computing Machinery*, 15:8–36.

Salton G, McGill MJ (1983). *Introduction to Modern Information Retrieval*. McGraw-Hill.

Sanderson M (1994). Word sense disambiguation and information retrieval. In *Proc. of the 17th International Conference on Research and Development in Information Retrieval*.

Sanfilippo A, Calzolari N, Ananiadou S, Gaizauskas R, Saint-Dizier P, Vossen P, Alonge A, Bel N, Bontcheva K, Bouillon P, Buitelaar P, Busa F, Harley A, Kamel M, Nogues MM, Montemagni S, Diez-Orzas P, Pianesi F, Pirrelli V, Segond F, Sjögreen C, Stevenson M, Gronostaj MT, Montserrat MV, Zampolli A (1999). Preliminary Recommendations on Lexical Semantic Encoding. Final Report EAGLES LE3-4244, The EAGLES (Expert Advisory Group on Language Engineering Standards) Lexicon Interest Group. http://www.ilc.cnr.it/EAGLES96/EAGLESLE.PDF.

Sathi A, Fox MS, Greenberg M (1985). Representation of Activity Knowledge for Project Management. *IEEE Transactions on pattern analysis and machine intelligence*, PAMI-7(5):531–552.

Schaerf A (1994). *Query answering in Concept-Based Knowledge Representation Systems: Algorithms, Complexity and Semantic Issues*. PhD thesis, Dipartimento di Informatica e Sistemistica. Università di Roma 'La Sapienza'.

Schatz B (1995). Building the Interspace: The Illinois Digital Library Project. *Comunications of the ACM*, 38(4).

Scherer D, Brennan C (2001). Exploring Oracle Text Basics. *Oracle Magazine*, March/April. http://www.oracle.com/oramag/index.html.

Sherwood LE (1998). Standards for access to museum content: practical solutions or technical chimeras? *Computer Standards & Interfaces*, 20:111–115.

Sheth A (1999). *Interoperating Geographic Information Systems*, chapter Changing Focus on Interoperability in Information Systems: from System, Syntax, Structure to Semantics, pp 5–29. Kluwer Academic Publishers, Boston.

Somers   R   (1997).   *Framework   Introduction   and   Guide*.   Federal   Geographic   Data   Committee   (FGDC).   http://www.fgdc.gov/framework/frameworkintroguide.

Stanford University School of Medicine (2004). The Protégé Project. http://protege.stanford.edu/.

Sussna M (1993). Word sense disambiguation for free-text indexing using a massive semantic network. In *Proc. of the Second International Conference on Information and Knowledge Management (CIKM-93)*, Arlington, Virginia.

Swoboda W, Nikolai FKR, Kazakos W, Nyhuis D, Rousselle H (1999). The UDK Approach: the 4th Generation of an Environmental Data Catalogue Introduced in Austria and Germany. In *Proc. of 3rd IEEE META-DATA Conference*, Bethesda, Maryland.

Teng Y (2000). Use of XML for Web-Based Query Processing of Geospatial Data. Master's thesis, University of New Brunswick.

Thompson HS, Beech D, Maloney M, Mendelsohn N, (eds) (2001). *XML Schema Part 1: Structures*. W3C, W3C Recommendation 2 May 2001. http://www.w3.org/TR/2001/REC-xmlschema-1-20010502/.

Tärnlund S (1977). Horn clause computability. *BIT*, 17:215–226.

Turner KJ, (ed) (1993). *Using Formal Description Techniques - An Introduction to ESTELLE, LOTOS and SDL*. Wiley, New York, USA.

Ullman JD (1989). *Principles of Database and Knowledge-Base Systems. Vols. 1 & 2*. Computer Science Press.

UNESCO (1995). *UNESCO Thesaurus: A Structured List of Descriptors for Indexing and Retrieving Literature in the Fields of Education, Science, Social and Human Science, Culture, Communication and Information*. United Nations Educational, Scientific and Cultural Organization (UNESCO) Publishing, Paris. http://www.ulcc.ac.uk/unesco/.

U.S. Federal Register (1994). Executive Order 12906. Coordinating Geographic Data Acquisition and Access: the National Spatial Data Infrastructure (U.S.). *The April 13,1994, Edition of the Federal Register*, 59(71):17671–17674.

U.S. Library of Congress (1996).   Z39.50 Profile for access to digital collections.   Final draft for review, Z39.50 Maintenance Agency. http://lcweb.loc.gov/z3950/agency/profiles/collections.html.

U.S. Library of Congress (1998). Encoded Archival Description (EAD) DTD. http://lcweb.loc.gov/ead/.

U.S. Library of Congress (2004a). *Library of Congress Subject Headings*. 27th edition.

U.S. Library of Congress (2004b). *MARC standards*. Network Development and MARC Standards office. http://www.loc.gov/marc/.

U.S. National Research Council (1999).   *Distributed Geolibraries: Spatial Information Resources. Summary of a Workshop: Panel on Distributed Geolibraries*.   National Academy Press, Washington, D.C. http://www.nap.edu/html/geolibraries.

Uzuner O (1998). Word Sense Disambiguation Applied to Information Retrieval. Master's thesis, Massachusetts Institute of Technology.

van Harmelen F, Patel-Schneider PF, Horrocks I, (eds) (2001). *DAML+OIL ontology markup language*.   DAML.Org, Reference description. http://www.daml.org/2001/03/reference.

van Rijsbergen CJ (1979). *Information Retrieval*, chapter Evaluation, pp 112–140. Dept. of Computer Science, University of Glasgow, 2nd edition.

Visser PRS, Jones DM, Bench-Capon TJM, Shave MJR (1997). An Analysis of Ontological Mismatches: Heterogeneity versus Interoperability. In *AAAI 1997 Spring Symposium on Ontological Engineering*, Stanford, USA.

Visser U, Stuckenschmidt H, Schuster G, Vögele T (2002).   Ontologies for geographic information processing. *Computers & Geosciences*, 28.

Vogel RL, Northcutt RT (1999). Integrating Inventory-Level Metadata into a Directory: An Implementation for Data Collections. In *Proc. of 3rd IEEE META-DATA Conference*, Bethesda, Maryland.

Voorhees EM (1993a). On Expanding Query Vectors with Lexically Related Words. In *Text REtrieval Conference*, pp 223–232.

Voorhees EM (1993b). Using WordNet to disambiguate Word Senses for Text Retrieval. In *SIGIR '93, Proc. 16th annual international ACM SIGIR conf. on Research and Development in Information Retrieval*, pp 171–180.

Voorhees EM (2002). Overview of TREC 2002. In *Proc. of 11th Text REtrieval Conference (TREC-11)*, Maryland.

Vretanos PA, (ed) (2001). *Filter Encoding Implementation Specification, Version 1.0.0*. Open Geospatial Consortium Inc (Open GIS Consortium Inc), OpenGIS Project Document OGC 02-059.

Vretanos PA, (ed) (2002). *Web Feature Server Implementation Specification. Version 1.0.0*. Open Geospatial Consortium Inc (Open GIS Consortium Inc), OpenGIS project document OGC 02-058.

W3C (2004a).   The Extensible Stylesheet Language Family (XSL). http://www.w3.org/Style/XSL/.

W3C (2004b). The Semantic Web Activity. http://www.w3.org/2001/sw/.

Weißenberg N, Gartmann R (2003). Ontology Architecture for Semantic Geo Services for Olympia 2008. In *Proc. of the GI-days 2003 conference*, pp 267–283, Münster (Germany).

Wiederhold G (1995). Digital Libraries, Value and Productivity. *Comunications of the ACM*, 38(4).

Wilensky R (1995). UC Berkeley's Digital Library Project. *Comunications of the ACM*, 38(4).

Winston M, Chaffin R, Herrmann DJ (1987). A Taxonomy of Part-Whole Relationships. *Cognitive Science*, 11:417–444.

Wood D (1995). *Computer Science Today: Recent Trends and Developments*, volume 1000 of *Lecture Notes in Computer Science*, chapter Standard Generalized Markup Language: Mathematical and Philosophical Issues, pp 344–365. Springer.

Woodley M (2000). Crosswalks: the Path to Universal Access? Getty Research Institute, Getty Standards and Digital Resource Management. http://www.getty.edu/research/institute/standards/intrometadata/2_articles/woodley/index.html.

Yarowsky D (1992). Word-sense disambiguation using statistical models of Roget's categories trained on large corpora. In *Proc. of the 14th International Conference on Computational Linguistics (Coling'92)*, pp 454–460, Nantes, France.

Zarazaga FJ, Bañares JA, Bernabé MA, Gould M, Muro-Medrano PR (2000a). La Infraestructura Nacional de Información Geográfica desde la Perspectiva de Bibliotecas Digitales Distribuidas. In *Proc. of the I Jornadas de Bibliotecas Digitales*, pp 163–172, Valladolid, España.

Zarazaga FJ, López R, Nogueras J, Cantán O, Álvarez P, Muro-Medrano PR (2000b). Cataloguing and recovering distributed geospatial data, a Java approach to build the OpenGIS Catalog Services. In *Proc. of the 6th European Commission GI&GIS Workshop, The Spatial Information Society - Shaping the Future*, Lyon, France.

Zarazaga FJ, López R, Nogueras J, Cantán O, Álvarez P, Muro-Medrano PR (2000c). First Steps to Set Up Java Components for the OpenGIS Catalog Services and its Software Infrastructure. In *Proc. of the 3rd AGILE Conference on Geographic Information Science*, pp 168–170, Helsinki/Espoo, Finland.

Zarazaga FJ, Torres MP, Nogueras-Iso J, Lacasta J, Cantán O (2003). Integrating geographic and non-geographic data search services using metadata crosswalks. In *Proc. of the 9th European Commission GI&GIS Workshop, ESDI: Serving the User*, A Coruña, Spain.

Zarazaga-Soria FJ (2000). *Una aproximación a la mejora de la reusabilidad de código C++ basada en metainformación del modelo de objetos*. PhD thesis, University of Zaragoza.

Zarazaga-Soria FJ, Lacasta J, Nogueras-Iso J, Torres MP, Muro-Medrano PR (2003a). A Java Tool for Creating ISO/FGDC Geographic Metadata. In *Proc. of the GI-days 2003 conference*, pp 17–30, Münster, Germany.

Zarazaga-Soria FJ, Nogueras-Iso J, Béjar R, Muro-Medrano PR (2004). Political aspects of Spatial Data Infrastructures. In Traunmüller R, (ed), *3rd International Conference on Electronic Government (EGOV'04)*, volume 3183 of *Lecture Notes in Computer Science*, pp 392–395, Zaragoza, Spain.

Zarazaga-Soria FJ, Nogueras-Iso J, Ford M (2003b). Dublin Core Spatial Application Profile. CWA 14858, CEN/ISSS Workshop - Metadata for Multimedia Information - Dublin Core.

Zarazaga-Soria FJ, Nogueras-Iso J, Ford M (2003c). Guidance material for mapping between Dublin Core and ISO in the Geograpic Information domain. CWA 14856, CEN/ISSS Workshop - Metadata for Multimedia Information - Dublin Core.

Zarazaga-Soria FJ, Nogueras-Iso J, Ford M (2003d). Mapping between Dublin Core and IS 19115, "Geographic Information Metadata". CWA 14857, CEN/ISSS Workshop - Metadata for Multimedia Information - Dublin Core.

Zipf GK (1945). The meaning-frequency relationship of words. *Journal of General Psychology*, 3:251–256.

Index

Abstract Data Type (ADT), 207
ACT ONE, 207
Aggregations, XII, 31, 48, 181
  aggregate, 36
  aggregate information, 37
  aggregation information, 35
  component information, 35
  container, 35
  initiative, 36
  mosaic, 31
  other aggregates, 36
  product, 31
  series, 31, 36
Alexandria Digital Library, 45
  Alexandria Digital Earth Prototype, 6
  Alexandria Digital Library Feature Type Thesaurus, 190
  Alexandria Digital Library Thesaurus Protocol, 238
Algebraic specification, 207
ANSI/NISO
  ANSI/NISO Z39.19, 133
  ANSI/NISO Z39.50, 19
  ANSI/NISO Z39.85, 10

BiblioTech, 237
BNF, 100, 116
Brokers, 9
BT, 134

C++, 26, 49
Capabilities, 17, 22, 40
Catalog, XI, 55, 171

catalog services
  access services, 19
  discovery services, 19
  management services, 19
  geographic data catalog, 19
  services Catalog, 21
CatMDEdit, 170, 178, 234
CEN, 10, 101, 120
  CEN/ISSS MMI/DC, 120
  CEN/TC 287, 14
  prENV 12657, 14, 97
CERES/NBII Thesaurus Partnership project, 238
CHIN, 96
Classic, 25
Clio, 98
CLIPS, 26
CLOS, 26
Collections, XII, 31, 181
  generation of statistics, 70
  mosaic, 31
  multiple-type collections, 32
  nested collections, 32, 198
  product, 31
  series, 31
  single-type collections, 32
Common Logic, 26
Compiler, 108
Conflicts
  domain conflicts, 99
  meta-data conflicts, 99, 102
  naming conflicts, 99
  structural conflicts, 99

Connexion, 98
CORBA, 19, 89
CORC, 98
Corpsmet95, 233
Cross-language, 168
Crosswalks, 91, 173, 199, 224
CSDGM, 11, 14, 97, 115, 173, 231
    Content Standard for Digital
        Geospatial Metadata, 14
    Remote Sensing extensions, 33, 35

DAML
    DAML+OIL, 17, 27
    Web Service Ontology (DAML-S), 17
DARPA, 27
Darwin Core, 96
Data mining, 76
Data Networks, 3
Data providers, 3
Databases, 3
    deductive database systems, 77
    Extensional Database, 77
    Intensional Database, 77
Datalog, 77
DC, XII, 7, 10, 35, 96, 115, 129, 173
    ANSI/NISO Z39.85, 10
    Dublin Core application profile, 12,
        101
    Dublin Core Metadata Initiative
        (DCMI), XII
    ISO 15836, 10
    Metadata Element Set, 10
    Qualified DC, 123
    Qualified DC metadata, 11
    Simple DC, 123
    Simple DC metadata, 11
Description Logics, 25, 72
Dialog Bluesheets, 43
Digital Library, X, 2, 5, 42, 129
Disambiguation, 131
    multilingual disambiguation, 168, 203
    semantic disambiguation, 131, 136,
        189
    thesaurus disambiguation, XIV, 131,
        189, 200
    training corpus, 149
    Word Sense Disambiguation (WSD),
        136
Discovery services, XII

E-commerce, 4
EAD, 44, 96
EIONET, 98
End-users, 3
Enraemed, 234
ESMI, 14
ESRI
    ArcCatalog, 235
    ArcIMS, 235
    ArcInfo, 22
    ArcView, 22
    SHAPE, 89, 184
ETeMII, 97
European Commission, X

F(rame)-Logic, 26, 57, 95
    F-Logic/Florid, 79
    LoPiX, 79
    XPathLog, 79
Facet, 24, 58
Feature
    geographic feature, 71
FGDC, 14
    Federal Geographic Data Committe,
        14
    FGDC Clearinghouse, 98
Frame, 24, 58
    facet, 24, 49, 58
    frame-based languages, 24, 49, 60
    Frame-Slot-Facet, 58
    slot, 24, 58

GCMD, 46, 190
    GCMD DIF, 96
GDDD, 14
GELOS, 14, 98
GEMET, 142, 149, 190
Geolibrary, XII, 5, 42
GI, VII
    Geographic Information, VII
    producers, VIII
GILS, 96
GIS, VIII, 1
    Geographic Information System,
        VIII, 1
    vendors, VIII
GML, 20, 71, 184
    Geographic Markup Language, 20

Grammar, 108
  Context Free Grammar, 108
  Extended Context Free Grammar, 109
  Regular Tree Grammar, 109
GSDI, 2, 8
  GSDI hierarchy, 8

Heterogeneity, XII, 89, 197
  heterogeneous databases, 99
  semantic heterogeneity, 99
HTML, 12, 22, 89, 106, 178
  HTML META tags, 12, 97
HTTP, 17, 19, 174

IDEE, 192
IGN, 31, 49, 192
Inference, 34, 48, 55, 57, 62
Information retrieval, XII, 128, 176, 200
  indexing, 155
  information retrieval model, 152
    vector-space retrieval model, 154
  Information Retrieval System (IRS), 132
  training corpus, 158
INSPIRE, X
Institutional arrangements, 3
Interoperability, XII, XIV, 9, 89, 199
  metadata interoperability, 90
  semantic interoperability, 89
  semantic mapping, 99, 102
  syntactic interoperability, 89
ISearch, 159
ISO
  ISO 11179, 101
  ISO 15836, 10
  ISO 19115, XII, 7, 11, 14, 17, 36, 61, 97, 115, 173, 231
  ISO 19119, 17, 21
  ISO 19139, 61, 97, 192
  ISO 23950, 19
  ISO 2788, 129, 133
  ISO 5964, 133, 238
  ISO TC/211, 14, 18

Java, 26, 61, 113, 144, 175, 234
JESS, 26

KIF, 26

KL-One, 25
Knowledge
  Knowledge Base System, 57
  Knowledge Discovery, 76
  Knowledge Representation, 2, 23, 47
  Metadata Knowledge Base, XIII, 57

LaClef, 14
LDOCE, 138
Legacy systems, 9
Lexico, 237
Library of Congress, 10
  Library of Congress Subject Headings, 129
LIMBER, 238
LISP, 26
Logic
  First order logic, 77
  Horn clause, 77
LOTOS, 207

M3Cat, 37, 233
MADAME, 97
MARC, 10, 90
  MAchine-Readable Cataloguing, 10
  MARC 21, 10
  MARC21, 129
  USMARC, 10
Metadata, XI, 2, 3, 9
  automatic generation of metadata, 65
  legacy metadata, XII
  metadata and capabilities, 22, 40
  metadata corpus, 158
  metadata crosswalks, XIV, 91
  metadata editors, 178, 231
  metadata element, 210
  metadata extensions, 16
  metadata inference, 65
  metadata interoperability, 90
  metadata maintenance, 8
  metadata profile, 12, 16, 101
  metadata record, XIII, 207
  metadata standards, XII, XIV, 10
  service metadata, 17
Metadata distinctions
  coincident metadata, 53, 215
  collection level metadata, 45
  collection-specific metadata, 53, 208
  common metadata, 208

contextual metadata, 45, 52
discovery metadata, 16
exploitation metadata, 16
exploration metadata, 16
inherent metadata, 45, 52, 215
item level metadata, 46
source content summary, 44
source metadata attributes, 44
Metadata Knowledge Base, XIII, 55,
    57, 169, 198, 215
MetaLite, 232
MetaMaker, 233
MetaManager, 233
MetaNet, 95
MIGRA, 14, 173, 192
Mosaic, 31
Multilingual, 168, 203
MultiTes, 237

NSDI, X, 2
NT, 134

OCLC, 98
OGC, X, 18
    Catalog specification, 19, 21, 171
    Common Query Language, 21, 73
    Filter Encoding Specification, 19
    Open Geospatial Consortium, X
    Open GIS Consortium, X
    Services Catalog, 22
    Web Feature Server, 20, 22
    Web Map Server, 20, 40, 86
OIL, 27
OKBC, 24
OntoEdit, 26
Ontolingua, 24
Ontology, 2, 23, 57, 94, 131
    ontology and spatial data infrastruc-
        tures, 27
    ontology languages, 24
    ontology tools, 24, 239
OWL, 27

Parser, 108
Policies, 3
Political aspects, 8
Precision-recall curves, 165
Product, 31
Profile

metadata profile, 12, 101
Prolog, 24, 77
Protégé, 24, 57, 60
Public administrations, VIII

Query answering, 72
    intelligent query answering, 72

RDF, 12, 94, 122
    RDF Vocabulary Description
        Language, 12, 94
    RDF-Schema, 12, 27, 94
    RDF/XML, 122, 238
    RDFS, 12, 94
Refutation, 77
Relations, 47
    AggregationRelation, 83
        MultipleTypeRelation, 84
        SingleTypeRelation, 83
        SpatialRelation, 84
        SpatioTemporalRelation, 84
        TemporalRelation, 84
        ThematicRelation, 85
    epistemological layer
        aggregation, 48
        classification, 47
        defaults, 47
        elaboration, 47
        individuation, 48
        revision, 48, 87
    format, 87
    high-level aggregation, 88
    part-whole relations, 48
        Component/Integral Objects, 84
        Component/Integral-Object, 48
        Feature/Activity, 49
        Member/Collection, 48, 84
        Place/Area, 49
        Portion/Mass, 49, 83
        Stuff/Object, 49
    relation roles
        conformsTo, 35
        hasFormat, 35, 87
        hasPart, 35
        hasVersion, 35, 86
        isFormatOf, 35, 87
        isPartOf, 35
        isReferencedBy, 35
        isReplacedBy, 35, 87

isRequiredBy, 35
isVersionOf, 35, 86
references, 35
replaces, 35, 87
requires, 35
revision, 87
version, 86
Roget's International Thesaurus, 138
RPN, 21, 73
RSLP, 44
RT, 134
Rules, 57, 62
    deduction rules, 77
    production rules, 35

SDI, 1
    GSDI, 2
    NSDI, X, 2
    ontology and spatial data infrastruc-
        tures, 27
    Spatial Data Infrastructure, X, 1
Semantic disambiguation, XIV, 136, 189
Semantic mapping, 99
Semantic Networks, 24, 47
Semantic Web, 23, 27, 94
Series, 31
    spatial series, 31
    temporal series, 31
    tile, 31
Sesame, 27
SGML, 44
SHOE, 95
SKOS, 239
Slot, 24, 58
SMART, 159
SOAP, 17, 21
Social aspects, 8
SQL, 19, 46, 55, 73, 77
Standards, 3
    standardization processes, 8
STARTS, 44, 52
SWAD-Europe, 239
Synaptica, 238

Technologies, 3
TermTree, 237
Thesaurus, XIV, 128, 133
    monolingual thesaurus, 129, 133
    multilingual thesaurus, 133

thesaurus disambiguation, XIV, 131,
    142, 189, 200
thesaurus exchange formats, 238
thesaurus management, 185
thesaurus relations, 133
thesaurus tools, 236
Tile, 31
TREC, 159
Tree
    parse tree, 108
    source tree, 110
    syntax tree, 108
    target tree, 110

UDDI, 17, 21
UDK, 14, 98, 130
UF, 134
UML, 36, 89, 116
USE, 134
USGS, 159
UTM, 15

VRA, 96

W3C, 10, 12, 17
    WWW Consortium, VIII, 10
Web mapping, 4
Web Services, 17
    Web Services Description Language
        (WSDL), 17
    Web Services Architecture, 17, 21
WebChoir, 237
Word Sense Disambiguation (WSD),
    136
WordNet, 131, 135, 186, 200
    EuroWordNet, 168, 203

XML, 10, 17, 19, 61, 106
    DOM, 109
    DTD, 10, 61, 100, 106
    XML parser, 109
    XML-Schema, 10, 61, 100, 106
    XPath, 73, 106, 107
    XSL, 10, 92, 106, 225
    XSLT, 106
XSB, 77

Z39.50, 10, 44, 73, 159
    BER, 19
    XER, 19